Robert Lang

Shear Viscosities from Kubo Formalism

Robert Lang

Shear Viscosities from Kubo Formalism

in a Large-Nc Nambu-Jona-Lasinio Model

Südwestdeutscher Verlag für Hochschulschriften

Impressum / Imprint
Bibliografische Information der Deutschen Nationalbibliothek: Die Deutsche Nationalbibliothek verzeichnet diese Publikation in der Deutschen Nationalbibliografie; detaillierte bibliografische Daten sind im Internet über http://dnb.d-nb.de abrufbar.
Alle in diesem Buch genannten Marken und Produktnamen unterliegen warenzeichen-, marken- oder patentrechtlichem Schutz bzw. sind Warenzeichen oder eingetragene Warenzeichen der jeweiligen Inhaber. Die Wiedergabe von Marken, Produktnamen, Gebrauchsnamen, Handelsnamen, Warenbezeichnungen u.s.w. in diesem Werk berechtigt auch ohne besondere Kennzeichnung nicht zu der Annahme, dass solche Namen im Sinne der Warenzeichen- und Markenschutzgesetzgebung als frei zu betrachten wären und daher von jedermann benutzt werden dürften.

Bibliographic information published by the Deutsche Nationalbibliothek: The Deutsche Nationalbibliothek lists this publication in the Deutsche Nationalbibliografie; detailed bibliographic data are available in the Internet at http://dnb.d-nb.de.
Any brand names and product names mentioned in this book are subject to trademark, brand or patent protection and are trademarks or registered trademarks of their respective holders. The use of brand names, product names, common names, trade names, product descriptions etc. even without a particular marking in this work is in no way to be construed to mean that such names may be regarded as unrestricted in respect of trademark and brand protection legislation and could thus be used by anyone.

Coverbild / Cover image: www.ingimage.com

Verlag / Publisher:
Südwestdeutscher Verlag für Hochschulschriften
ist ein Imprint der / is a trademark of
OmniScriptum GmbH & Co. KG
Heinrich-Böcking-Str. 6-8, 66121 Saarbrücken, Deutschland / Germany
Email: info@svh-verlag.de

Herstellung: siehe letzte Seite /
Printed at: see last page
ISBN: 978-3-8381-5130-4

Zugl. / Approved by: München, TU, Diss., 2015

Copyright © 2015 OmniScriptum GmbH & Co. KG
Alle Rechte vorbehalten. / All rights reserved. Saarbrücken 2015

Contents

1. **Motivation**   7

2. **Quantum Chromodynamics and the Quark-Gluon Plasma**   9
   - 2.1. Quantum Chromodynamics – a symmetry-guided overview   9
     - 2.1.1. QCD Lagrangian and its main features   9
     - 2.1.2. PCAC and low-energy theorems of QCD   13
     - 2.1.3. Large-N_c extension of QCD and key aspects of AdS/CFT correspondence   15
   - 2.2. Heavy-ion collisions and the quark-gluon plasma   20
     - 2.2.1. Experimental facilities and a standard model of heavy-ion collisions   20
     - 2.2.2. Hydrodynamic description of the quark-gluon plasma   24
   - 2.3. Kubo formalism   28
     - 2.3.1. Transport coefficients from linear-response theory   28
     - 2.3.2. Ladder-diagram resummation in the Kubo formalism   29

3. **The Nambu–Jona-Lasinio model**   33
   - 3.1. General N_f, N_c-NJL Lagrangian in the chiral limit   33
   - 3.2. The two-flavor NJL model   41
   - 3.3. Large-N_c analysis for the NJL model   44
     - 3.3.1. Topology of connected QCD vertices   44
     - 3.3.2. Generating functional of the NJL model   47
   - 3.4. Gap equation and thermal constituent-quark masses   51
   - 3.5. Bethe-Salpeter equation and thermal meson masses   54
     - 3.5.1. Meson masses and mesonic spectral functions   54
     - 3.5.2. Quark-meson coupling beyond the pole-mass approximation   58
   - 3.6. Pion decay constant and low-energy theorems   61
   - 3.7. Attractive diquark channels   62

4. **Microscopic theory of the shear viscosity**   67
   - 4.1. Shear viscosity from Kubo formalism   67
   - 4.2. Parameter study of the shear viscosity   73
     - 4.2.1. Analytical results for a constant spectral width   74
     - 4.2.2. Numerical approach to momentum-dependent spectral widths   75
     - 4.2.3. Cutoff dependence   76
     - 4.2.4. Perturbative aspects and ladder-diagram resummation   78
     - 4.2.5. Effects of thermal quark masses on the shear viscosity   79
   - 4.3. Kinetic theory   80

5. **Mesonic fluctuations in the quark sector**   85
   - 5.1. Ambiguous analytical continuation from discrete data   85
   - 5.2. Quark self-energy from mesonic fluctuations   87
     - 5.2.1. On-shell contributions   87
     - 5.2.2. Off-shell contributions   94

Contents

    5.3. Vacuum fluctuations and the cloudy bag model 100

6. The ratio η/s in the NJL model   105
    6.1. Entropy density in a large-N_c expansion 105
    6.2. Kubo formalism for the Dirac self-energy 108
    6.3. Results for the shear viscosity and the ratio η/s 110

7. Summary and Conclusion   115

A. Appendix   117
    A.1. Group-theoretic details of SU(N) 117
    A.2. Matsubara formalism 120
        A.2.1. Review of propagators 120
        A.2.2. Master formulas 121
    A.3. Spectral representation of propagators 123
    A.4. Analytical properties of the quark-meson coupling 125
    A.5. List of symbols 127

1. Motivation

> "Before the Standard Model we could not go back further than 200,000 years after the Big Bang. Today, especially since QCD simplifies at high energy, we can extrapolate to very early times, where nucleons melt and quarks and gluons are liberated to form a quark-gluon plasma."[Gro05]
>
> David Gross, Nobel Lecture 2004

The world we are experiencing on a daily basis is dominated by electromagnetism and gravity. In our low-temperature world at $T \approx 300$ K the strong force which is described by Quantum Chromodynamics (QCD) is restricted to microscopic scales due to the phenomenon of confinement: quarks and gluons cannot be observed freely but they form hadrons such as nucleons and mesons. Nature realizes a deconfined state of QCD matter only under extreme conditions present in the early universe with $T \gtrsim 0.2$ GeV $\approx 2 \cdot 10^{12}$ K, or perhaps in the center of (cold) compact stars. Since the year 2000 the creation of such an extreme state of hot matter has become possible also on Earth using collider experiments with gold, copper or lead ions. They have been carried out first at the Relativistic Heavy Ion Collider (RHIC) of the Brookhaven National Laboratory (BNL). Nowadays, collisions at even higher center-of-mass energies are pursued at the Large Hadron Collider (LHC) of the European Organization for Nuclear Research (CERN).

In such ultra-relativistic heavy-ion collisions a quark-gluon plasma is produced. It is understood to be a highly correlated system behaving like an almost-perfect fluid with small viscosities. It is remarkable that the measurement of particle-flow patterns in comparison to results from hydrodynamic simulations allow the extraction of the shear viscosity to entropy ratio η/s. This ratio has been found to be close to the benchmark $1/4\pi$ calculated from principles based on the AdS/CFT correspondence. In particular, for the quark-gluon plasma it is smaller than for any other fluid studied so far, sketched in Fig. 1.1. The AdS/CFT benchmark refers to a perfect fluid with infinitely strong correlations. Small values of the shear viscosity indicate indeed a highly correlated system of quarks and gluons. In this thesis we aim to calculate the shear viscosity $\eta(T,\mu)$ as function of temperature and quark chemical potential within the two-flavor Nambu–Jona-Lasinio (NJL) model. Such a T- dependent viscosity can serve as input for hydrodynamic simulations and can therefore help improving the understanding of the quark-gluon plasma.

The outline of this thesis is as follows: In **Chapter 2** we review the basics of our work starting with symmetries of QCD and the AdS/CFT correspondence. We examine also the synergy between experimental flow-results and hydrodynamic simulations. The Kubo formalism which allows the derivation of the shear viscosity from a microscopic quantum field theory is briefly reviewed, too. In **Chapter 3** we introduce the NJL model and discuss its large-N_c properties. It is shown how standard results such as the gap equation and the Bethe-Salpeter equation describing mesonic (soft) modes can be derived within this formalism. The general Kubo formalism for the shear viscosity becomes more specific when applying it to the NJL model as it is done in the first part of **Chapter 4**. In its second part we investigate essential features of the shear viscosity in a parameter study assuming that the shear viscosity is represented by one single momentum-dependent spectral width. We introduce a suitable numerical approximation scheme and investigate how the three-momentum cutoff and the thermal constituent-quark affect the shear viscosity. The kinetic approach is reviewed and parallels and differences with respect to

1. Motivation

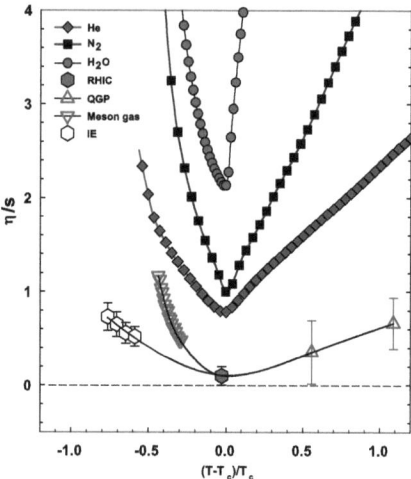

Figure 1.1.: Ratio η/s as function of temperature in the vicinity of the respective critical/crossover temperatures T_c for different hydrodynamic systems. The value corresponding to the quark-gluon plasma produced in ultra-relativistic heavy-ion collisions has been found to be very small and close to the AdS/CFT benchmark $1/4\pi$. Figure taken from [L+07].

the Kubo formalism are discussed. In **Chapter 5** we consider mesonic fluctuations contributing to the self-energy of both on-shell and off-shell constituent quarks. These fluctuations provide the dominant dissipative process contributing to the shear viscosity of the hot and dense plasma we are investigating. Furthermore, we compare our findings to well-known results from chiral (cloudy) bag models and show that they are consistent with our more general field-theoretical calculation. **Chapter 6** starts with the discussion how the entropy density can be derived within the large-N_c NJL model. Then we generalize the Kubo formalism used for the parameter study in order to incorporate the full Dirac structure of the quark-self energy calculated from mesonic fluctuations. Our results for the shear viscosity and the ratio η/s are shown as functions of temperature and quark chemical potential. We also compare them to lattice results and other approaches. Finally, we summarize our most important findings in **Chapter 7**. The closing **Appendix A** contains some technical details and general reviews for techniques we have used in this thesis.

2. Quantum Chromodynamics and the Quark-Gluon Plasma

> *"It is wrong to think that the task of physics is to find out how nature is.*
> *Physics concerns what we can say about nature..."[McE01]*
>
> Niels Bohr

In this introductory chapter we review the three topics which form the basis of this work. First, the symmetry patterns and main features of Quantum Chromodynamics (QCD) are discussed. We introduce the large-N_c extension of QCD and investigate how it affects its structure. Second, heavy-ion collisions are reviewed and it is illustrated how hydrodynamic simulations are used to extract the ratio η/s (shear viscosity to entropy density) from anisotropic-flow measurements. The main goal of this work is to calculate the temperature dependence of this ratio using the Nambu–Jona-Lasinio (NJL) model. Therefore, third, we introduce the Kubo formalism which provides a microscopic description of dissipative parameters such as shear viscosity. We discuss how the Kubo formula can be derived from linear response theory and pinpoint its complexity arising from a diagrammatical treatment.

2.1. Quantum Chromodynamics – a symmetry-guided overview

2.1.1. QCD Lagrangian and its main features

The strong interaction of quarks and gluons is described by Quantum Chromodynamics given in terms of the Lagrangian[1]

$$\mathcal{L}_{\text{QCD}} = \bar{\psi}\left(\mathrm{i}\slashed{D} - \hat{m}\right)\psi - \frac{1}{2}\text{Tr}\left(G_{\mu\nu}G^{\mu\nu}\right), \tag{2.1}$$

gauged by the color symmetry $\text{SU}(3)_c$ with the physical number of colors $N_c = 3$. The fermion field $\psi = (\psi_1, \ldots, \psi_{N_f})^T$ collects the $N_f = 6$ quark flavors with their additional color structure $\psi_i = (\psi_i^r, \psi_i^g, \psi_i^b)^T$. The mass matrix is diagonal in flavor space, $\hat{m} = \text{diag}(m_1, \ldots, m_{N_f})$, and trivial in color space due to the gauge symmetry. In the *isospin limit* all quarks feature the same mass, $\hat{m} = m_0 \mathbb{1}_{N_f \times N_f}$. In the *chiral limit* these current-quark masses are sent to zero: $m_0 \to 0$. Whereas the chiral limit is a good approximation for $N_f = 2$ flavors only, the isospin limit is sometimes used also for the three-flavor case.

In the QCD Lagrangian (2.1) the *covariant derivative* is denoted by

$$D_\mu = \partial_\mu - \mathrm{i}g_{\text{QCD}} A_\mu^a T^a, \tag{2.2}$$

with g_{QCD} the fundamental coupling strength of QCD, A_μ^a denoting the eight gluon (vector boson) fields, and T^a the (infinitesimal) generators of the gauge group[2]. The gluonic field-

---
[1] For reviewing QCD we refer to standard textbooks, e.g. [PS95, Wei99, ESW03], and partly to [Lan10].
[2] Some group-theoretic details about the Lie group SU(N) are discussed and summarized in the Appendix A.1.

2. Quantum Chromodynamics and the Quark-Gluon Plasma

	QCD Lagrangian			QCD vacuum
global flavor symmetry	full QCD	isospin limit	chiral limit	$\langle \bar\psi\psi \rangle \neq 0$
$SU(N_f)_L \times SU(N_f)_R$	✗	✗	✓	✗
$SU(N_f)_V$	✗	✓	✓	✓
$U(1)_V$	✓	✓	✓	✓
$U(1)_A$	✗	✗	classical	✗

Table 2.1.: Summary of global flavor symmetries of the QCD Lagrangian (2.1) and the vacuum state. ✗ denotes an absent symmetry and ✓ denotes a present symmetry, cf. the discussion in the text.

strength tensor is
$$\begin{aligned} G_{\mu\nu} &= G_{\mu\nu}^a T^a = ig_{\text{QCD}}^{-1}[D_\mu, D_\nu], \\ G_{\mu\nu}^a &= \partial_\mu A_\nu^a - \partial_\nu A_\mu^a + g_{\text{QCD}} f^{abc} A_\mu^b A_\nu^c, \end{aligned} \quad (2.3)$$

where the gauge fields are assumed to be smooth functions on Minkowski space, $A_\mu^a \in \mathcal{C}^2[\mathbb{M}]$. The non-Abelian structure of QCD is encoded in the non-vanishing structure constants $f_{abc} \neq 0$. They are cyclic and totally antisymmetric and lead to three-gluon and four-gluon vertices. Apart from the gauge symmetry QCD is also invariant under Poincaré transformations and features the discrete $\mathcal{C} \times \mathcal{P} \times \mathcal{T}$ symmetries.

In flavor space there are (under certain approximations) additional global symmetries: at the classical level and in the chiral limit the QCD Lagrangian (2.1) is invariant under

$$U(N_f)_L \times U(N_f)_R = SU(N_f)_L \times SU(N_f)_R \times U(1)_L \times U(1)_R. \quad (2.4)$$

As we discuss later in more detail (cf. Eq. (3.62) and the related instanton discussion), the axialvector symmetry is anomalously broken: $U(1)_L \times U(1)_R \mapsto U(1)_V$. This can be understood from the non-trivial transformation behavior of the path-integral measure $\int \mathcal{D}\psi \mathcal{D}\bar\psi$ under $\psi \mapsto \exp(-i\gamma_5 \alpha)\psi$ [Fuj79]. It explains the rather large η'-meson mass of 958 MeV.

In this work, the *chiral symmetry* $SU(N_f)_L \times SU(N_f)_R$ and its breaking mechanisms are of fundamental importance. We introduce left- and right-handed quark fields by

$$\psi_{L/R} = \frac{1}{2}(1 \mp \gamma_5)\psi, \quad (2.5)$$

which allow us to rewrite the QCD Lagrangian (2.1) as

$$\mathcal{L}_{\text{QCD}} = \bar\psi_L (i\slashed{D}) \psi_L + \bar\psi_R (i\slashed{D}) \psi_R - (\bar\psi_L \hat{m} \psi_R + \bar\psi_R \hat{m} \psi_L) - \frac{1}{2}\text{Tr}(G_{\mu\nu}G^{\mu\nu}). \quad (2.6)$$

It is obvious that in the chiral limit left- and right-handed quark fields can be transformed independently. A finite mass term breaks this symmetry, because it mixes ψ_L and ψ_R. However, in the isospin limit the chiral symmetry is broken explicitly to the remnant $SU(N_f)_V$ symmetry where left- and right-handed fields transform only simultaneously via $\psi \mapsto \exp(-i\alpha_a T_a)\psi$.

In Table 2.1 all global flavor symmetries are summarized. On the level of the Lagrangian, symmetries are broken explicitly, whereas the vacuum state breaks symmetries spontaneously. Having in the chiral limit also the anomalous breaking of $U(1)_A$ symmetry (denoted by "classical"), QCD features all possible symmetry-breaking mechanisms present in a relativistic quantum field theory. We also mention the possibility to add both a $\mathcal{P}$ and $\mathcal{C} \times \mathcal{P}$ (and therefore

2.1. Quantum Chromodynamics – a symmetry-guided overview

also $\mathcal{T}$) violating θ-term to the QCD Lagrangian (2.1):

$$\mathcal{L}_\theta = -\frac{\theta}{32\pi^2} G_{\mu\nu} \widetilde{G}^{\mu\nu}, \qquad (2.7)$$

with $\widetilde{G}^{\mu\nu} = \frac{1}{2}\epsilon^{\mu\nu\alpha\beta}G_{\alpha\beta}$ denoting the dual gluonic field-strength tensor and $\theta \in \mathbb{C}$ some free parameter. Despite the fact that $\mathcal{L}_\theta$ is a total derivative (cf. again Eq. (3.62) and the related instanton discussion), it gives rise to an electric dipole moment of the neutron (nEDM) [CDVVW79, Dar00]:

$$|d_\mathrm{n}| \approx g_{\pi\mathrm{NN}} \frac{0.04\,|\theta|}{4\pi^2 m_\mathrm{N}} \ln\frac{m_\mathrm{N}}{m_\pi} e = 2.7 \cdot 10^{-5} |\theta|\, e\,\mathrm{MeV}^{-1} = 5.4 \cdot 10^{-16} |\theta|\, e\,\mathrm{cm}, \qquad (2.8)$$

where $m_\mathrm{N} = 939$ MeV is the neutron (nucleon) mass and $g_{\pi\mathrm{NN}} = 13.2 \pm 0.1$ denotes the pion-nucleon coupling constant. All experimental searches so far are consistent with a vanishing nEDM and one can state an upper limit at $|d_\mathrm{n}| < 2.9 \cdot 10^{-26}\, e\,\mathrm{cm}$ (with 90% confidence level) [B+06]. Therefore, the QCD θ-term is small, $|\theta| \lesssim 10^{-10}$, leading to the so-called strong CP problem which is discussed elsewhere [Wil78, CL06]. In most applications and investigations of QCD the θ-term is set to zero.

Apart from the current-quark masses there is no scale introduced by the QCD Lagrangian. Therefore, on the classical level, QCD is scale invariant in its chiral limit. This scale invariance is only one of the aspects of conformal symmetry which is anomalously broken by quantum effects leading to a non-vanishing beta function[3]:

$$\beta(a_\mathrm{s}) = \mu \frac{\partial a_\mathrm{s}(\mu)}{\partial \mu} = -2 \sum_{n=1}^{\infty} \beta_{n-1} a_\mathrm{s}^{n+1}(\mu), \qquad (2.9)$$

with μ being the renormalization scale and a_s denoting the reduced strong fine-structure constant:

$$a_\mathrm{s} = \frac{\alpha_\mathrm{s}}{4\pi} = \frac{g_\mathrm{QCD}^2(\mu)}{16\pi^2}. \qquad (2.10)$$

In the SU(N_c) case, the coefficients read in the $\overline{\mathrm{MS}}$ renormalization scheme [vRVL97, Cza05]:

$$\beta_0 = \frac{11}{3}N_\mathrm{c} - \frac{2}{3}N_\mathrm{f},$$
$$\beta_1 = \frac{34}{3}N_\mathrm{c}^2 - \frac{10}{3}N_\mathrm{c}N_\mathrm{f} - \frac{N_\mathrm{c}^2-1}{N_\mathrm{c}}N_\mathrm{f},$$
$$\beta_2 = \frac{2857}{54}N_\mathrm{c}^3 + \frac{(N_\mathrm{c}^2-1)^2}{4N_\mathrm{c}^2}N_\mathrm{f} - \frac{205}{36}(N_\mathrm{c}^2-1)N_\mathrm{f} - \frac{1415}{54}N_\mathrm{c}^2 N_\mathrm{f} + \frac{11}{18}\frac{N_\mathrm{c}^2-1}{N_\mathrm{c}}N_\mathrm{f}^2 + \frac{79}{54}N_\mathrm{c}N_\mathrm{f}^2. \qquad (2.11)$$

At leading order this leads to[4]:

$$\beta(a_\mathrm{s}) = -\beta_0 \frac{\alpha_\mathrm{s}^2}{2\pi} \quad \Rightarrow \quad \alpha_\mathrm{s}(\mu) = \frac{\alpha_\mathrm{s}(\mu_0)}{1 + \frac{\beta_0}{4\pi}\alpha_\mathrm{s}(\mu_0) \ln\left(\frac{\mu^2}{\mu_0^2}\right)}. \qquad (2.12)$$

For $\beta_0 > 0$, i.e. for $N_\mathrm{f} < 11 N_\mathrm{c}/2 = 17$, the beta function is negative and *asymptotic freedom* is realized meaning that g_QCD features small values at high energy scales allowing for a perturbative treatment. This is fulfilled for the physical values, $N_\mathrm{c} = 3$ and $N_\mathrm{f} = 6$, but also in the large-N_c

---
[3]Note that there are several equivalent definitions of the beta function using different variables, e.g. $\beta(\alpha_\mathrm{s}) = 4\pi\beta(a_\mathrm{s})$ or $\beta(g_\mathrm{QCD}) = \frac{2\pi}{g_\mathrm{QCD}}\beta(a_\mathrm{s})$.
[4]Later in Section 3.3 we use the beta function to introduce the large-N_c scaling in the NJL model.

2. Quantum Chromodynamics and the Quark-Gluon Plasma

limit. It is known that only non-Abelian Yang-Mills Lagrangians can feature asymptotic freedom [WG73, CG73, Pol73, Gro05]. In 2004 Gross, Politzer, and Wilczek were awarded the Nobel Prize in Physics "for the discovery of asymptotic freedom in the theory of the strong interaction".

From the anomalous breaking of conformal symmetry expressed by the beta function an intrinsic energy scale emerges: Λ_{QCD}. It is defined from the pole $\alpha_s^{-1}(\Lambda_{\text{QCD}}) = 0$. From Eq. (2.12) one finds at leading order:

$$\Lambda_{\text{QCD}} = \mu_0 \exp\left(-\frac{2\pi}{\beta_0 \alpha_s(\mu_0)}\right). \tag{2.13}$$

For four active quark flavors at the renormalization scale $\mu_0 = 2$ GeV and $\alpha_s(\mu_0) \approx 1/3$ it evaluates to $\Lambda_{\text{QCD}} \approx 0.2$ GeV. The fact that the coupling α_s appears in Eq. (2.13) in the exponent's denominator expresses the non-perturbative nature of Λ_{QCD}. This scale separates the perturbative from the non-perturbative sector of QCD. It also separates the light hadrons (e.g. pions) from the more energetic states (e.g. nucleons) of the physical spectrum.

For low energies the strong coupling becomes large indicating the phenomenon if *confinement*. We emphasize that the beta function is derived using Feynman diagrams which is a perturbative technique, therefore conclusions about large α_s are beyond the applicability of this approach. Confinement is still not fully understood and there are several aspects of confinement which can be approached in different ways. One possible approach is provided from a closed Wilson line in imaginary time, the *Polyakov loop*:

$$L(\boldsymbol{x}) = \mathcal{P} \exp\left(i \int_0^\beta d\tau \, A_4(\boldsymbol{x}, \tau)\right), \tag{2.14}$$

where $\mathcal{P}$ denotes the path-ordering symbol, $\beta = 1/T$ is the inverse temperature and A_4 is the fourth component of the gluon field field $A_\mu = A_\mu^a T^a$. Introducing also the renormalized Polyakov loop,

$$\Phi(\boldsymbol{x}) = \frac{1}{N_c} \text{Tr}_c \, L(\boldsymbol{x}), \tag{2.15}$$

one can show that [MS81]:

$$\langle \Phi(\boldsymbol{x}) \Phi^\dagger(\boldsymbol{y}) \rangle_\beta = e^{-\beta \mathcal{F}_{q\bar{q}}(\boldsymbol{x}-\boldsymbol{y})}, \tag{2.16}$$

where $\mathcal{F}_{q\bar{q}}(\boldsymbol{x}-\boldsymbol{y})$ denotes the free energy of two static color sources q and $\bar{q}$ with spatial separation $\boldsymbol{r} = \boldsymbol{x} - \boldsymbol{y}$. Sending one color source to infinity, correlations between the two sources vanish and the thermal expectation value $\langle \cdot \rangle_\beta$ in Eq. (2.16) factorizes. This allows to relate the (thermal expectation value of the renormalized) Polyakov loop to the free energy of a single quark:

$$\langle \Phi \rangle_\beta = e^{-\frac{\beta}{2} \mathcal{F}_q^\infty}. \tag{2.17}$$

It can be seen that for a divergent free energy of a single quark one has $\langle \Phi \rangle_\beta = 0$. This is interpreted as confinement. In the deconfined phase one has $\langle \Phi \rangle_\beta \approx 1$. We note that strictly speaking this is only true in the pure gauge case without quarks, but the Polyakov loop is used as an order parameter for the deconfinement transition also in the matter case. Additionally, the Polyakov loop does not describe confinement in a sense that quarks are spatially clustered. It only ensures the suppression of colored configurations, denoted as *statistical confinement*.

Calculating the potential between two static color sources q and $\bar{q}$ using lattice QCD provides a more fundamental indication for confinement [B+00]. The lattice potential can be fitted by

$$V(r) = -\frac{e}{r} + \sigma r + \text{const.} \tag{2.18}$$

The *string tension* is calculated to be $\sqrt{\sigma} \approx 450$ MeV, cf. the review [Bal01] for instance. For large distances the linear part of the potential dominates and indicates confinement in the closer meaning: when separating two color sources too much, the spontaneous creation of a quark-antiquark pair becomes energetically favored, hence the spatial distance between two color sources is bounded from above. However, we mention that confinement is still not fully understood and subject of intense investigations.

2.1.2. PCAC and low-energy theorems of QCD

Modeling QCD by substituting its Lagrangian by some simpler Lagrangian modeling its symmetry pattern is a very fruitful and commonly used strategy to tackle, in particular, non-perturbative aspects of the strong interaction. In the low-temperature region one can use effective theories such as chiral perturbation theory (χPT), a systematic approach based on the chiral effective field theory guided by the (approximate) chiral symmetry of QCD, cf. [Sch02] for a reviewing introduction. Instead of an effective field theory, we will use in this work a model approach to QCD: the Nambu–Jona-Lasinio (NJL) model that will be introduced in detail in Chapter 3. However, when modeling QCD, one needs to guarantee low-energy theorems to be satisfied, such as the Gell-Mann-Oakes-Renner (GOR) relation and the Goldberger-Treiman (GT) relation [LK96, TW01, CL06]. They are based on the *partially conserved axial current* (PCAC) hypothesis and can be derived using only basic current-algebra techniques. Later in this thesis, we will discuss these relations again in Section 3.6, when demonstrating their validity within the NJL model.

From the QCD Lagrangian (2.1) the vector and axialvector current are defined as ($N_\mathrm{f} = 3$)

$$V_a^\mu(x) = \bar{\psi}(x)\gamma^\mu \frac{\lambda_a}{2}\psi(x) \,, \\ A_a^\mu(x) = \bar{\psi}(x)\gamma^\mu\gamma_5 \frac{\lambda_a}{2}\psi(x) \,, \quad (2.19)$$

where $\lambda_a = 2T_a$ denote the Gell-Mann matrices, $a = 1, \ldots, 8$. From these currents the (conserved) vector and axialvector charges follow:

$$Q_a^\mathrm{V} = \int \mathrm{d}^3x\, V_a^0(x) = \int \mathrm{d}^3x\, \psi^\dagger(x)\frac{\lambda_a}{2}\psi(x) \,, \\ Q_a^\mathrm{A} = \int \mathrm{d}^3x\, A_a^0(x) = \int \mathrm{d}^3x\, \psi^\dagger(x)\gamma_5\frac{\lambda_a}{2}\psi(x) \,. \quad (2.20)$$

We use for any Dirac structures Γ_1, Γ_2 and flavor structures F_1, F_2 the following identity,

$$[\Gamma_1 F_1, \Gamma_2 F_2] = \frac{1}{2}\{\Gamma_1, \Gamma_2\}[F_1, F_2] + \frac{1}{2}[\Gamma_1, \Gamma_2]\{F_1, F_2\} \,, \quad (2.21)$$

leading to the general equal-time commutator ($x_0 = y_0 = t$) [Sch02]:

$$\left[\psi^\dagger(x)\Gamma_1 F_1 \psi(x), \psi^\dagger(y)\Gamma_2 F_2 \psi(y)\right] = \delta^{(3)}(\boldsymbol{x}-\boldsymbol{y})\left(\psi^\dagger(x)\Gamma_1\Gamma_2 F_1 F_2 \psi(y) - \psi^\dagger(y)F_2 F_1 \Gamma_2 \Gamma_1 \psi(x)\right). \quad (2.22)$$

From this a straightforward calculation yields:

$$\left[Q_a^\mathrm{V}, \bar{\psi}(x)\lambda_b\psi(x)\right] = if_{abc}\bar{\psi}(x)\lambda_c\psi(x) \,, \\ \left[Q_a^\mathrm{A}, \bar{\psi}(x)\gamma_5\lambda_b\psi(x)\right] = -\bar{\psi}(x)\left(\frac{2}{3}\delta_{ab} + d_{abc}\lambda_c\right)\psi(x) \,, \quad (2.23)$$

where we have used the fundamental representation of the Lie algebra SU(3), hence $\xi(\mathrm{F}) = 2/3$

2. Quantum Chromodynamics and the Quark-Gluon Plasma

in the three-flavor case. For details we refer to the Appendix A.1.

The pion decay constant f_π is introduced by the following matrix elements which describes the vacuum annihilation of a pion through the axialvector current[5]:

$$\langle 0|A_a^\mu(x)|\pi_b(p)\rangle = \mathrm{i}p^\mu f_\pi \delta_{ab} \mathrm{e}^{-\mathrm{i}p \cdot x} , \qquad (2.24)$$

where the right-hand side is just a parametrization of the left-hand side due to its Lorentz structure. This definition refers to the vacuum case, i.e. $T = 0$ and $\mu = 0$. In order to avoid a parity doubling of the mesonic octet ($N_\mathrm{f} = 3$) in the low-energy spectrum, one chooses the Nambu-Goldstone realization of chiral symmetry, i.e. $Q_a^\mathrm{A}|0\rangle \neq 0$, cf. the discussion of the NJL phase diagram in Chapter 3.4. One finds

$$\langle 0|Q_a^\mathrm{A}(t=0)|\pi_b(p)\rangle = \int \mathrm{d}^3x\, \langle 0|A_a^0(x)|\pi_b(p)\rangle = \mathrm{i}E_p f_\pi \delta_{ab}(2\pi)^3 \delta^{(3)}(\boldsymbol{p}) , \qquad (2.25)$$

meaning that $Q_a^\mathrm{A}|0\rangle$ *contains* one-pion states (with zero momenta). The PCAC hypothesis states that the spectrum of Q_a^A acting on the vacuum, is *dominated* by one-pion states forming a complete set:

$$\int \frac{\mathrm{d}^3p}{2E_p(2\pi)^3}|\pi_a(p)\rangle\langle\pi_a(p)| = \mathbb{1} . \qquad (2.26)$$

We note for clarity, that $Q_a^\mathrm{A}|0\rangle$ actually contains all states with the corresponding quantum numbers, e.g. three-particle states

$$\int \frac{\mathrm{d}^3p\,\mathrm{d}^3q\,\mathrm{d}^3k}{8E_p E_q E_k (2\pi)^9}|\pi_a(p)\otimes\pi_b(q)\otimes\pi_c(k)\rangle\langle\pi_a(p)\otimes\pi_b(q)\otimes\pi_c(k)| . \qquad (2.27)$$

From the PCAC hypothesis two important low-energy theorems can be deduced: the Gell-Mann–Oakes-Renner (GOR) relation and the Goldberger-Treiman (GT) relation. We focus on the GOR relation which will be derived in the following. Taking only the explicit chiral symmetry breaking through the mass term $\hat{m} = \mathrm{diag}(m_\mathrm{u}, m_\mathrm{d}, m_\mathrm{s})$ into account we have:

$$\begin{aligned}\partial_\mu A_1^\mu(x) &= \mathrm{i}\bar\psi(x)\{\mathrm{diag}(m_\mathrm{u}, m_\mathrm{d}, m_\mathrm{s}), \frac{\lambda_1}{2}\}\psi(x) = \\ &= (m_\mathrm{u}+m_\mathrm{d})\bar\psi(x)\mathrm{i}\gamma_5\frac{\lambda_1}{2}\psi(x) .\end{aligned} \qquad (2.28)$$

Commuting with Q_1^A and sandwiching between vacuum states we find

$$\begin{aligned}\langle 0|[Q_1^\mathrm{A},\partial_\mu A_1^\mu]|0\rangle &= (m_\mathrm{u}+m_\mathrm{d})\mathrm{i}\langle[Q_1^\mathrm{A},\bar\psi(x)\gamma_5\frac{\lambda_1}{2}\psi(x)]\rangle = \\ &= -\frac{\mathrm{i}}{2}(m_\mathrm{u}+m_\mathrm{d})\langle\bar u u + \bar d d\rangle ,\end{aligned} \qquad (2.29)$$

where we have used Eq. (2.23) for $a = b = 1$, so $d_{118} = 1/\sqrt{3}$ and $\lambda_8 = 1/\sqrt{3}\,\mathrm{diag}(1,1,-2)$. Therefore, we get

$$\langle 0|[Q_1^\mathrm{A},\bar\psi(x)\gamma_5\lambda_1\psi(x)]|0\rangle = -\left(\langle\bar u u\rangle + \langle\bar d d\rangle\right) . \qquad (2.30)$$

We mention that this equation can be used for defining the chiral condensate $\langle\bar\psi\psi\rangle$, but at this stage we postpone a more detailed discussion to Section 3.4. Inserting now the PCAC hypothesis, i.e. a full set of one-pion states, into both terms of the commutator of Eq. (2.29),

---
[5]From the decay $\pi^- \to \mu^- \bar\nu(\gamma)$ its experimental value is determined to $f_\pi = 91.92 \pm 0.02 \pm 0.14$ MeV. The first (smaller) error is due to uncertainties of $|V_{ud}| = 0.97425(22)$ whereas the second (larger) error is due to higher-order corrections [O+14].

its left-hand side reads:

$$\int \mathrm{d}^3 p \left(\underbrace{\langle 0|Q_1^A|\pi_a(p)\rangle}_{\mathrm{i}E_p f_\pi \delta_{1a}(2\pi)^3 \delta^{(3)}(\boldsymbol{p})} \underbrace{\langle \pi_a(p)|\partial_\mu A_1^\mu|0\rangle}_{m_\pi^2 f_\pi \delta_{1a}} - \underbrace{\langle 0|\partial_\mu A_1^\mu|\pi_a(p)\rangle}_{m_\pi^2 f_\pi \delta_{1a}} \underbrace{\langle \pi_a(p)|Q_1^A|0\rangle}_{-\mathrm{i}E_p f_\pi \delta_{1a}(2\pi)^3 \delta^{(3)}(\boldsymbol{p})} \right) = 2\mathrm{i}(2\pi)^3 f_\pi^2 m_\pi^2 E_p \,, \tag{2.31}$$

where we have used $\mathrm{i}\partial_\mu p^\mu = m_\pi^2$. Integrating also the right-hand side of Eq. (2.29) over all three-momenta, only the Lorentz-covariant normalization $\int \mathrm{d}^3 p = 2E_p(2\pi)^3$ is introduced. We arrive at the GOR relation:

$$m_\pi^2 f_\pi^2 = -\frac{1}{2}(m_u + m_d)\left(\langle \bar{u}u\rangle + \langle \bar{d}d\rangle\right) \approx -m_0 \langle \bar{\psi}\psi\rangle \,. \tag{2.32}$$

This relation expresses a simple connection between quantities from explicit (m_0) and spontaneous ($m_\pi, \langle \bar{\psi}\psi\rangle$) symmetry breaking. As we have seen, its derivation is based only on fundamental QCD symmetries and their breaking patters.

2.1.3. Large-N_c extension of QCD and key aspects of AdS/CFT correspondence

The main difference between QED (Quantum Electrodynamics) and QCD is the gauge group which is Abelian for QED but non-Abelian for QCD. As we have reviewed above, this structure leads to specific features like asymptotic freedom and confinement. Especially the lack of a general perturbative technique applicable not only to the high-energy region of QCD is challenging and makes QCD far more complex.

In December 1973 G. 't Hooft came up with a new topological classification of (QCD) Feynman diagrams when describing interacting quarks: *planar diagrams* dominate for $N_\mathrm{c} \to \infty$ in a U(N_c) or SU(N_c) gauge group [tH74a]. In QCD, of course, the number of colors is fixed from experiment and not a free parameter. The textbook method to do so is considering the R-ratio from e^+e^- collisions [EJ91]:

$$R(s) = \frac{\sigma_\mathrm{tot}(e^+e^- \to \gamma^* \to hadrons)}{\sigma_\mathrm{tot}(e^+e^- \to \gamma^* \to \mu^+\mu^-)} = N_\mathrm{c} \sum_{f=1}^{N_f} q_f^2 \sqrt{1 - \frac{4m_f^2}{s}} \left(1 + \frac{2m_f^2}{s}\right)\left(1 + \frac{\alpha_\mathrm{s}}{4\pi} + \mathcal{O}(\alpha_\mathrm{s}^2)\right), \tag{2.33}$$

where $\sqrt{s}$ denotes the center-of-mass energy. We note that in the actual measurement the total hadronic cross section includes QED corrections from bremsstrahlung and vacuum-polarization effects which changes the details of how to compare the experimental data to $R(s)$ from a pure QCD calculation. However, when exploring the large-N_c generalization of QCD, one realizes from Eq. (2.12) that the strong coupling constant becomes small, $g_\mathrm{QCD} \to 0$, but the 't Hooft coupling λ becomes a (scale dependent) constant:

$$\lambda(\mu) = g_\mathrm{QCD}^2(\mu)N_\mathrm{c} = 4\pi\alpha_\mathrm{s}(\mu)N_\mathrm{c} \longrightarrow \frac{48\pi^2}{11\ln\left(\frac{\mu^2}{\mu_0^2}\right)} \quad \text{as } N_\mathrm{c} \to \infty \,. \tag{2.34}$$

This is the relevant coupling in a SU(N_c) Yang-Mills theory in its large-N_c limit. As pointed out by Witten, the expansion parameter $1/N_\mathrm{c}$ is non-obvious and there is no a priori reason why one should perform such an analysis:

> "The hope is that it may be possible to solve the theory [QCD] in the large N limit, and that the $N = 3$ theory may be qualitatively and quantitatively close to the large N limit." [Wit79]

For instance, asymptotic freedom as one of the main features of QCD is not qualitatively affected

2. Quantum Chromodynamics and the Quark-Gluon Plasma

by a large-N_c expansion. The leading-order condition $\beta_0 > 0$ is always fulfilled in this limit, independent of the number of quark flavors, cf. the coefficients in Eq. (2.11).

QCD is not an isolated theory but it is part of the Standard Model with the gauge group $SU(3)_c \otimes SU(2)_L \otimes U(1)_Y$. In a naive large-$N_c$ expansion one would simply substitute by

$$SU(N_c)_c \otimes SU(2)_L \otimes U(1)_Y \,. \tag{2.35}$$

Ensuring the renormalizability of the large-N_c extended Standard Model it is necessary to have all chiral anomalies canceled. As investigated in [CY97] there are triangle diagrams which do not vanish trivially:

$$U(1)_Y^3 \,,\quad U(1)_Y SU(2)_L^2 \,,\quad U(1)_Y SU(N_c)_c^2 \,. \tag{2.36}$$

Forcing these diagrams to vanish sets non-linear constraints on the $U(1)_Y$ hypercharges which are solved and discussed in [GM89, MRW90, GM90]. In conclusion the electroweak charges (in units of the elementary charge e) of the first quark and lepton generation are given by

$$(q_u, q_d, q_e, q_\nu) = \left(\frac{1+N_c}{2N_c}, \frac{1-N_c}{2N_c}, -1, 0\right). \tag{2.37}$$

We always have $q_u = q_d + 1$ but their actual value varies with the number of colors. For $N_c = 3$ one finds the standard values $q_u = 2/3$ and $q_d = -1/3$. In contrast, the lepton charges are not affected by N_c and keep their standard values. When increasing the number of colors one has to distinguish between an odd and even number series. Usually the large-N_c series is interpreted as $(N_c)_n = 2n + 1 \in (3, 5, 7, 9, \ldots)$ with $n \in \mathbb{N}$, where the proton consists of $n + 1$ up-quarks and n down-quarks ensuring the physical proton and neutron charge:

$$(n+1) \cdot \frac{1+N_c}{2N_c} + n \cdot \frac{1-N_c}{2N_c} = \frac{2n+1+N_c}{2N_c} = 1 \,. \tag{2.38}$$

The neutron consists of n up-quarks and $n+1$ down-quarks which reproduces its neutral electric charge as well:

$$n \cdot \frac{1+N_c}{2N_c} + (n+1) \cdot \frac{1-N_c}{2N_c} = \frac{2n+1-N_c}{2N_c} = 0 \,. \tag{2.39}$$

In the picture of valence quarks, the leading-order Fock state contains $(n+1) + n = N_c$ quarks, therefore both proton and neutron are ensured to be color neutral. This is true for an odd number of colors. Allowing also an even number of quarks, $(N_c)_n = 2n \in (2, 4, 6, 8, 10, \ldots)$ with $n \in \mathbb{N}$, leads to a totally different world since nucleons become bosons. In the literature especially the two-color case is studied intensely. Its thermodynamics is described by model approaches [RW04] or effective field theories [KST99, KST+00], which can interpret but also induce lattice studies [SSS01, NFH04, LP13]. Excluding the possibility of parity-breaking phases, two-color QCD is special for lattice calculations because the prohibitive sign problem is not present there [Fuk07]. In a two-color world, nucleons are just diquarks and the proton consists of one single up- and one single down-quark.[6]

Within the last few decades one highlighted aspect of $SU(N_c)$ Yang-Mills theories is their (con-

---

[6] We would like to mention one further possible large-N_c extensions of the Standard Model. From the observation that $R(s) \sim N_c$ one can criticize that in the naive extension (2.35) the hadronic processes are favored but leptonic ones are suppressed. As it is suggested in [Erd98], one can introduce a new global symmetry group $SU(N_c/3)_l$ to the Standard Model acting only on leptons which belong to the fundamental representation of this group:
$$SU(N_c)_c \otimes SU(N_c/3)_l \otimes SU(2)_L \otimes U(1)_Y \,.$$
Also this formalism uses the physical constraints from anomaly cancellation to fix the quark and lepton charges. Apart from the advantage of the leptonic sector featuring the same weight for large values of N_c, the resulting quark charges do not change with the number of colors.

jectured) duality[7] with string theories. The fact that planar diagrams provide the leading-order contribution to large-N_c QCD has its analogy in string theory: there, diagrams are organized in a loop expansion where the genus χ, counting the holes in diagrams, is the relevant topological parameter. Assuming a small string length l_s, the scattering of e.g. two closed strings is organized in an expansion in l_s^χ. This basic analogy was already pointed out 1974 by 't Hooft and in the meantime several non-trivial dualities have been found. In the following we refer mostly to the review [BAA+12] and highlight the correspondence between string theory on anti-de Sitter (AdS) space and some conformal field theory (CFT). This turned out to be relevant within the context of heavy-ion collisions where QCD processes are dominant.[8] The *strong form* of AdS/CFT correspondence reads:

> There is an exact duality between type IIB string theory on $AdS_5 \times S^5$ and four-dimensional maximally ($\mathcal{N}=4$) superconformal Yang-Mills theory.

Symmetry dictates that the considered Yang-Mills theory must feature additionally superconformal symmetry, i.e. it must be supersymmetric and conformal[9]. Besides the simple scale invariance, there are also special conformal transformations which extend the Poincaré group in total by five generators from $SO(1,3)$ to $SO(2,4)$. As a submanifold of this supergroup, the five-dimensional anti-de Sitter space is described by the metric

$$ds^2 = \underbrace{\frac{r^2}{R^2}\left(-dt^2 + d\boldsymbol{x}^2\right)}_{\text{4 dimensions}} + \underbrace{\frac{R^2}{r^2}dr^2}_{\text{1 dimension}} . \tag{2.40}$$

The constant curvature radius R of AdS_5 sets an intrinsic length scale for the bulk described by the coordinate r. The conformal boundary of AdS_5 is reached at $r = \infty$, where flat Minkowski space in $4+1$ dimensions is located. Copies of the sphere S^5 are affixed at each spacetime point ensuring to provide a ten-dimensional spacetime where string theory can live. It is known that the symmetry group of $AdS_5 \times S^5$ is the same as the superconformal group in $3+1$ spacetime dimensions [HLS75]. One can relate the closed string coupling g to the (running) coupling constant of the Yang-Mills theory by $g_{YM}^2 = 4\pi g$, cf. [BBS06] or any other standard string-theory textbook. In addition, one has

$$\frac{R^4}{\alpha'^2} = 4\pi g N_c = g_{YM}^2 N_c = \lambda , \tag{2.41}$$

with α' denoting the Regge slope. Historically, when string theory was actually constructed to explain the strong force, it was defined empirically from the hadron spectrum [ZSV04, AP09]: as it has been observed, both the squared meson and baryons masses, M^2, can be described easily by the Regge trajectory

$$J = \alpha' M^2 , \tag{2.42}$$

where J denotes the hadron spin. At least for mesons this simple form can be explained intuitively[10] by a fast rotating relativistic string: $M^2 = 2\pi\sigma J$ with $\sqrt{\sigma}$ being the string tension. By comparison the so-called QCD string is given by $\sigma = (2\pi\alpha')^{-1}$. Empirically, for the hadron case, the numerical value of the Regge slope is $\alpha' = \mathcal{O}(1 \text{ GeV}^{-2})$. However, the Regge slope in the context of AdS/CFT correspondence serves as some free model parameter.

---
[7]Two theories are called dual if the functional structures of the external sources s in the corresponding partition functions $Z[s] = \int \left(\prod_{\text{d.o.f.}} \mathcal{D}X\right) e^{-iS[X,s]}$ coincide.
[8]Of course, in heavy-ion collisions there are QED processes like photoproduction as well.
[9]We mention that in two dimensions the conformal transformations are given by the biholomorphic functions which is of crucial importance for the two-dimensional world sheet in string theory.
[10]We mention that quantum effects alter this form, introducing a non-vanishing intercept and a positive curvature. [ZSV04]

2. Quantum Chromodynamics and the Quark-Gluon Plasma

	Quantum Field Theory	String Theory
$\lambda \ll 1$	perturbative	full theory with large string fluctuations
$\lambda \gg 1$ ('t Hooft limit)	non-perturbative	supergravity with small string fluctuations

Table 2.2.: Limits of the 't Hooft coupling λ and its impact on QFT and string theory

Having a look back to the $SU(N_c)$ Yang-Mills quantum field theory, we realize that only for a small 't Hooft coupling a perturbative treatment is possible. In this regime the curvature R in AdS$_5$ space needs to be also small as it can be seen from Eq. (2.41). This means that string fluctuations (measured by the string length) become of similar size as the length scale of the bulk space: $l_s \approx R$. Under this condition a perturbative treatment of string theory, i.e. an expansion in l_s^χ as mentioned before, is not possible anymore. Therefore, it is not possible to have a perturbatively accessible quantum field theory and corresponding string theory at the same time. We summarize this discussion in table 2.2. As a consequence, the strong form of AdS/CFT correspondence is hard to prove. However, in the so-called *'t Hooft limit*, where the 't Hooft coupling becomes large, $\lambda \gg 1$, one has the *weak form* of AdS/CFT correspondence:

The large-λ, large-N_c limit of four-dimensional maximally ($\mathcal{N}=4$) superconformal Yang-Mills theory is dual to classical type IIB supergravity on AdS$_5 \times S^5$.

This duality is just one, out of several examples, that were proven by Maldacena in his celebrated paper [Mal99].

For this thesis applications based on the weak form of AdS/CFT duality are relevant: as we will explain in detail in Section 2.2 the ratio η/s is of crucial importance for the study of heavy-ion collisions, where η denotes the shear viscosity and s the entropy density. The Kubo formalism (cf. Section 2.3) provides one possible approach to transport coefficients like shear viscosity. Within a strongly coupled quantum field theory, its exact evaluation is almost impossible and one usually applies (more or less rough) approximation schemes. However, it is also possible to tackle this task using AdS/CFT correspondence which has been done first by Kovtun, Son and Starinets in the acclaimed paper [KSS05]. We note that a thermal medium (in particular the temperature T) is introduced to a quantum field theory usually by using the Matsubara formalism as it is known from standard textbooks [LB00, KG06]. The dual description of a thermal medium in anti-de Sitter space is a Schwarzschild black hole (SBH) which is described by the metric[11]

$$ds^2 = \frac{r^2}{R^2}\left(-f(r)dt^2 + d\boldsymbol{x}^2\right) + \frac{R^2}{r^2}\frac{dr^2}{f(r)}, \quad \text{with } f(r) = 1 - \left(\frac{r_0}{r}\right)^4. \tag{2.43}$$

In the limit $r \to r_0$ (approaching the black-hole horizon from flat space, i.e. $r > r_0$ and $0 < f(r) < 1$), the metric becomes singular which indicates the presence of a black hole[12]. As it is shown in [PSS01], the shear viscosity of a strongly-coupled supersymmetric Yang-Mills plasma reads

$$\eta(T) = \frac{\pi}{8} N_c^2 T^3. \tag{2.44}$$

---
[11] It is interesting to note that the scale factor $f(r)$ affects not only the bulk but also the time coordinate in the conformal boundary of AdS$_5$ space. This can be interpreted as analogy to the Matsubara formalism where also the temporal axis becomes affected by compactifying onto $[0, 1/T]$ implying the discrete spectrum of Matsubara frequencies.

[12] Since its horizon is translationally invariant one sometimes denotes it as black brane.

This result has been derived using AdS/CFT correspondence and the definition of η as Kubo correlator[13], calculating the absorption cross section of a graviton by a black brane in AdS_5 where the temperature T is the Hawking temperature of the metric (2.43):

$$T = \frac{r_0}{\pi R^2} \, . \tag{2.45}$$

The zero-temperature case is reached for $r_0 \to 0$, i.e. when pushing the black brane infinitely far away from the conformal boundary. Combining the result for $\eta(T)$ with the Bekenstein-Hawking entropy (density) [GKP96, Kov12],

$$s(T) = \frac{\pi^2}{2} N_c^2 T^3 \, , \tag{2.46}$$

one arrives at the famous ratio

$$\frac{\eta}{s} = \frac{1}{4\pi} \, . \tag{2.47}$$

This constant result is exact in $AdS_5 \times S^5$ quantum gravity, therefore, by applying the AdS/CFT correspondence, also for large-N_c superconformal $SU(N_c)$ Yang-Mills theory in the 't Hooft limit. For many reasons, QCD is far away from the considered Yang-Mills theory: of course, there are actually only three colors and QCD does not feature supersymmetry. Even in the chiral limit where no length scale is present in the Lagrangian (2.1), conformal symmetry is broken anomalously leading to $\beta(\alpha_s) \neq 0$ and the intrinsic energy scale Λ_{QCD}. However, as it has been also derived by Kovtun, Son and Starinets, the first-order correction to the ratio η/s in inverse powers of the 't Hooft coupling is positive:

$$\frac{\eta}{s} - \frac{1}{4\pi} = \frac{135\zeta(3)}{8(2\lambda)^{3/2}} > 0 \, . \tag{2.48}$$

From this observation the *viscosity bound conjecture* (KSS conjecture) has been drawn:

"Most quantum field theories do not have simple gravity duals. Is our result relevant in a broader setting? We speculate that the ratio η/s has a lower bound $\eta/s \geq 1/4\pi$ for all relativistic quantum field theories at finite temperature and zero chemical potential. The inequality is saturated by theories with gravity duals." [KSS05]

In fact, so far all experimentally accessible physical systems do respect this AdS/CFT bound. The more strongly the system is coupled the smaller one expects its corresponding ratio η/s. As we will discuss in detail in the next Section 2.2 the quark-gluon plasma produced in heavy-ion collisions features a remarkably small value of η/s. However, it is known that one can construct theories and models where this ratio is undershot, e.g. because the viscosity can be independent of the particle multiplicity but the entropy density is not [Coh07]. In an anisotropic plasma the KSS bound can be violated as well [Mam12, RS12].

In this thesis we are calculating the ratio η/s for the NJL model. Its non-perturbative nature can be explored applying a large-N_c expansion which is used as book-keeping method only and for any numerical result we always use the physical value $N_c = 3$. Although the NJL model is far away from being a quantum field theory possessing a quantum-gravity dual, it is, however, instructive to compare our results to the benchmark $\eta/s = 1/4\pi$ from AdS/CFT correspondence.

---

[13]We emphasize that this is the same footing our own analysis for $\eta(T,\mu)$ within the NJL model is based on.

2. Quantum Chromodynamics and the Quark-Gluon Plasma

2.2. Heavy-ion collisions and the quark-gluon plasma

In heavy-ion collisions matter can be studied under extreme conditions. There are two main facilities where gold, copper or lead ions are accelerated to ultra-relativistic velocities before they are collided: the Relativistic Heavy Ion Collider at the Brookhaven National Laboratory (RHIC@BNL) and the Large Hadron Collider at the European Organization for Nuclear Research (LHC@CERN). It is established that in both facilities a quark-gluon plasma (QGP) has been created as a strongly coupled system that behaves like an almost-perfect fluid with $\eta/s < 5/4\pi$, e.g. [Son13] and references therein. Until the shutdown of the LHC on 14 February 2013 there have been two runs with lead-lead collisions and one run with proton-lead collisions. It is already clear that the QGP produced at the LHC differs to some extend from the plasma produced at RHIC where the center-of-mass energy has been a few hundred GeV, whereas at LHC the TeV scale has been reached, cf. Table 2.3. In spring 2015 its upgrade is expected to be finished allowing then for collisions with $\sqrt{s_{NN}} = 5.1$ TeV. In this section we will describe the basic ideas of heavy-ion collisions and explain in detail how the plasma created at the LHC motivates this thesis asking for a better understanding of the temperature dependence of shear viscosity. In this section we refer mostly to the (review) articles [Ven10, HSS12, Sne11, Oll11, Son13] and the standard textbook [YHM08].

2.2.1. Experimental facilities and a standard model of heavy-ion collisions

A sketch of a heavy-ion collision is shown in Fig. 2.1 where the beam line is oriented along the z-axis. In the heavy-ion program of LHC, the heaviest known stable isotope, Pb^{208}, has been used which is a double-magic nucleus with spherical symmetry. Each collision event differs by its centrality class c which can be calculated from the impact parameter $\boldsymbol{b} = b\,\hat{\boldsymbol{e}}_x$:

$$c = \frac{\pi b^2}{\pi (2R_A)^2} \,, \tag{2.49}$$

where $R_A \approx 1.3$ fm $\cdot \sqrt[3]{A}$ is the radius of the nucleus. The lower the centrality class the more central is the collision. If there is no full geometric overlap between the two nuclei the collision is called peripheral. Nucleons that are not participating in the collision are called spectators: $N_{\text{spec}} = N - N_{\text{part}}$. In a central collision one has $N \approx N_{\text{part}} = 2A \approx 400$ participants. The impact parameter b is not directly observable and cannot be used for determining the centrality class. Instead, one measures the multiplicity of charged particles in an event, $dN_{\text{evt}}/dN_{\text{ch}}$, and averages over many events. A typical result is shown in Fig. 2.2 from the first elliptic-flow measurement at the LHC. Using the Glauber model, cf. [MRSS07] for a review, the final-state observable N_{ch} can be related to the impact parameter, b, and the number of participating

	active period	$\sqrt{s_{NN}}$	beam velocity
RHIC@BNL	2000 to present	200 GeV	$c - 13,200$ m s^{-1}
LHC@CERN	2009 to 2013	2.76 TeV	$c - 70$ m s^{-1}
LHC@CERN	starting in 2015	5.1 TeV	$c - 20$ m s^{-1}

Table 2.3.: Comparison of beam energies at RHIC and LHC for Au-Au and Pb-Pb collisions, respectively. The center-of-mass energy $\sqrt{s_{NN}}$ refers to the energy of each nucleon pair.

nucleons, N_part. This is based on the assumption that the underlying centrality classes and the impact parameter b are monotonically related to the particle multiplicity. The Glauber model parameterizes the density distribution of nuclei using a Woods-Saxon distribution:

$$\rho(r) = \frac{\rho_0}{1 + \exp\left(\frac{r-R_A}{a}\right)}, \qquad (2.50)$$

where we restrict to the case of spherical nuclei with three model parameters ρ_0, R_A, a, which have been determined from low-energy electron scattering experiments. In a Glauber Monte Carlo (GMC) simulation, nucleons are randomly distributed within the ion according to this distribution. A collision between two nucleons takes place if their distance satisfies the simple geometric condition $d < \sqrt{\sigma_\text{inel}^\text{pp}/\pi}$, where $\sigma_\text{inel}^\text{pp}$ denotes the inelastic proton-proton cross section, cf. Fig. 2.3 for its dependence on the center-of-mass energy $\sqrt{s}$. It is empirically know that the multiplicity of soft particles scales with the number of participating nucleons, $N_\text{ch} \approx \frac{2}{3} N_\text{soft} \sim N_\text{part}$, but the multiplicity of hard particles scales with the number of nucleon-nucleon collisions, $N_\text{hard} \sim N_\text{coll}$. From a simple geometric picture, N_coll depends on the volume of the interaction volume and its length in beam direction, $l_z \sim N_\text{part}^{1/3}$, hence $N_\text{coll} \sim N_\text{part}^{4/3}$. Considering for instance a central Au-Au collision at $\sqrt{s_\text{NN}} = 0.2$ GeV at RHIC [A+06] one has $N_\text{part} \approx 400$ but $N_\text{coll} \approx 0.4\, N_\text{part}^{4/3} \approx 1200$ with the cross section $\sigma_\text{inel}^\text{pp}\big|_{\sqrt{s_\text{NN}}=0.2\,\text{TeV}} = 42$ mb. The same collision at LHC energies would lead to an increase of N_coll by roughly 50% because the cross section increases to $\sigma_\text{inel}^\text{pp}\big|_{\sqrt{s_\text{NN}}=2.76\,\text{TeV}} = 64$ mb, resulting $N_\text{coll} \gtrsim 2000$.

Directly after the collision, a pre-equilibrium phase is created that starts to expand primarily one-dimensionally in the beam-line direction. At this stage, perturbative QCD techniques are applicable in principle. The color-glass condensate (CGC) model has been settled to describe this initial state successfully, cf. [ILM01, FILM02] and [GIJMV10] for a review. The CGC model is based on the assumption that the multiplicity of partons within the relativistic heavy ion is huge. This means that every single parton carries only a small fraction of the total ion energy. Since gluons dominate the parton-distribution functions at small Bjorken-x, the two colliding ions at ultra-relativistic energies can be described as a dense condensate of gauge color sources interacting perturbatively with each other.

As it is sketched in the Bjorken spacetime picture shown in Fig. 2.4, the thermalization converts the pre-equilibrium state after $\tau_0 \approx 1$ fm into the quark-gluon plasma where *local*

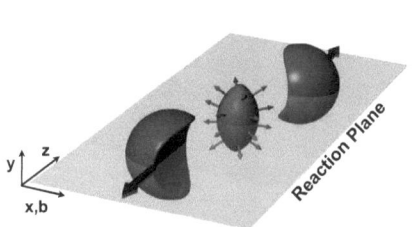

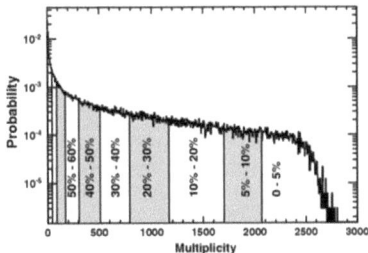

Figure 2.1.: Schematic heavy-ion collision with spectators and interaction volume in almond shape. Figure taken from [Sne11].

Figure 2.2.: Per-event charged particle multiplicity $N_\text{ev}^{-1}\,\text{d}N_\text{evt}/\text{d}N_\text{ch}$ averaged over $4.5 \cdot 10^4$ Pb-Pb collisions at ALICE. Figure taken from [A+10].

2. Quantum Chromodynamics and the Quark-Gluon Plasma

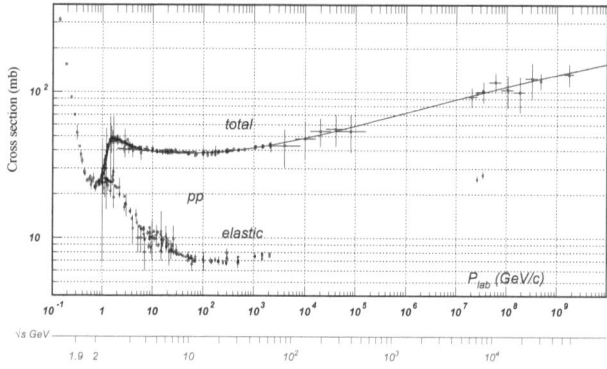

Figure 2.3.: Proton-proton cross sections as function of the center-of-mass energy. $\sigma_{\text{inel}}^{\text{pp}}$ is the difference between the total and elastic cross section. For low energies Coulomb repulsion suppresses the cross section, at high energies it scales approximatively as $\sigma \sim \log^2 s$. Figure taken from [O+14].

thermal equilibrium[14] is reached. As discussed in the next Section 2.2.2, the QGP can be described using viscous relativistic hydrodynamics. The QGP expands three-dimensionally and evolves towards hadronization where first chemical freeze-out and then thermal freeze-out takes place, $\tau_c < \tau_t$. For $\tau > \tau_c$ the numbers of each particle species (pions, nucleons, kaons, etc.) stay constant but the particles are still in local thermal equilibrium and their kinetic distributions are coupled. After the thermal freeze-out, $\tau > \tau_t$, the kinetic equilibrium is no longer maintained and the hadrons will be finally detected with some momentum and angular distribution.

In the detector one can measure the momentum-distribution of the particles for each hadron species. It can be expanded in a Fourier series introducing the flow coefficients v_n:

$$E \frac{\mathrm{d}^3 N}{\mathrm{d}^3 p} = \frac{1}{2\pi} \frac{\mathrm{d}^2 N}{p_T \, \mathrm{d}p_T \, \mathrm{d}Y} \left(1 + 2 \sum_{n=1}^{\infty} v_n \cos\left[n(\varphi - \Psi_{\text{RP}})\right] \right), \quad (2.51)$$

where E is the particle energy, φ the azimuthal angle, $p_T = |\boldsymbol{p}_T|$ the transverse momentum, $p^\mu = (p_0, \boldsymbol{p}_T, p_z)$, and Y denotes the *rapidity* of the observed particle. It is defined as

$$Y = \frac{1}{2} \ln \frac{E + p_z}{E - p_z} \approx -\ln \tan \frac{\theta}{2} = \eta_{\text{PS}}, \quad (2.52)$$

and simplifies in the (ultra) relativistic case to the *pseudorapidity*: $\lim_{|\boldsymbol{p}| \gg m} Y = \eta_{\text{PS}}$, where θ denotes the detection angle measured with respect to the beam line. Hence, particles with vanishing pseudorapidity have escaped the interaction volume perpendicularly to the beam line, whereas particles with large pseudorapidity can be found in the forward and backward detectors[15]. In Eq. (2.51) we have denoted the reaction-plane angle by Ψ_{RP} as it is shown in Fig. 2.1.

---

[14] A system is said to be in local thermal equilibrium if it is possible to divide the entire system into smaller cells where thermodynamic quantities like temperature, entropy and pressure can be defined. These quantities may differ for different cells, but they are approximately constant within one cell. In order to define first-order dissipative parameters as shear or bulk viscosity the system needs to be in local equilibrium.

[15] For the four detectors at RHIC (BRAHMS, PHENIX, PHOBOS and STAR) the pseudorapidity range that can be covered is roughly $|\eta_{\text{PS}}| < 5$, the same is true for ATLAS and CMS at CERN. ALICE, in contrast, is more restricted. There, only $|\eta_{\text{PS}}| < 1$ is used for flow-related measurement.

2.2. Heavy-ion collisions and the quark-gluon plasma

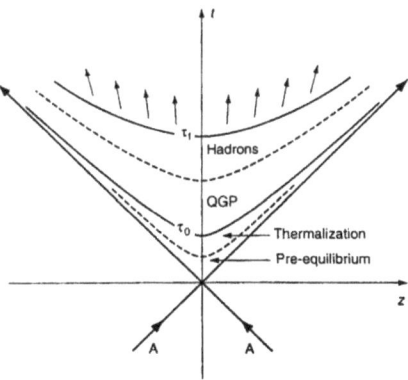

Figure 2.4.: Bjorken spacetime picture of a heavy-ion collision. Figure taken from [YHM08].

This angle varies from event to event and is not directly observable. As a consequence, the Fourier coefficients, v_n, cannot be measured directly:

$$v_n(p_\mathrm{T}, Y) = \langle \cos\left[n(\varphi - \Psi_\mathrm{RP})\right]\rangle \,, \tag{2.53}$$

where the brackets $\langle \cdot \rangle$ denote the particle average summed over all events. Assuming the simplest geometry, i.e. $\Psi_\mathrm{RP} = 0$, the first two flow coefficients read:

$$\begin{aligned} v_1 &= \langle \cos\varphi \rangle = \frac{\langle p_x \rangle}{\langle p_\mathrm{T} \rangle} \,, \\ v_2 &= \langle \cos 2\varphi \rangle = \frac{\langle p_x^2 \rangle - \langle p_y^2 \rangle}{\langle p_\mathrm{T}^2 \rangle} \,. \end{aligned} \tag{2.54}$$

One calls v_1 the *directed flow* and v_2 the *elliptic flow*. In general, the collective expansion of the QGP has been denoted as *flow*. For our purposes v_2 is most important because the ratio η/s can be extracted from its measurement as we will discuss in the next Section 2.2.2. The elliptic flow measures the anisotropy in the momentum distribution in the transverse plane. In non-central collisions, this anisotropy can be explained from the almond shape of the interaction volume. The initial spatial anisotropies,

$$\epsilon_n = \frac{\langle r^n \cos(n\varphi) \rangle}{\langle r^2 \rangle^{n/2}} \,, \tag{2.55}$$

are transferred into the anisotropies in the momentum distribution. The elliptic flow v_2 is mainly induced by the eccentricity ϵ_2, but in general there are also influences from $\epsilon_{m\neq n}$ to v_n. Also non-flow effects like jets, i.e. effects which are not due the collective expansion of the QGP, or initial fluctuations affect the flow coefficients v_n, cf. for instance [OPV09, GGLO12].

As we have mentioned, the coefficients v_n cannot be measured directly. An indirect approach is given from pair-particle $(2k)$ correlations, cf. [BDO01, MS03] or [Sne11] for a more general review. One introduces two-particle $(k=1)$ or four-particle $(k=2)$ cumulants via

$$\begin{aligned} c_n\{2\} &= \langle\langle \mathrm{e}^{\mathrm{i}n(\varphi_1-\varphi_2)} \rangle\rangle \,, \\ c_n\{4\} &= \langle\langle \mathrm{e}^{\mathrm{i}n(\varphi_1+\varphi_2-\varphi_3-\varphi_4)} \rangle\rangle - 2c_n^2\{2\} \,, \end{aligned} \tag{2.56}$$

2. Quantum Chromodynamics and the Quark-Gluon Plasma

where the double brackets $\langle\langle \cdot \rangle\rangle$ denote an average first over all particles in one event, and then over all events. The cumulants $c_n\{2k\}$ are observables, because only differences of azimuthal angles appear and the reaction plane angle drops out:

$$\Delta\varphi_{ij} = \varphi_i - \varphi_j = (\varphi_i - \Psi_{\text{RP}}) - (\varphi_j - \Psi_{\text{RP}}) . \tag{2.57}$$

Note that because of the reflection symmetry with respect to the reaction plane one has[16]

$$v_n = \langle \cos n(\varphi - \Psi_{\text{RP}}) \rangle = \langle e^{in(\varphi - \Psi_{\text{RP}})} \rangle , \tag{2.58}$$

and no sine term contributes. One finds therefore

$$\begin{aligned} c_2\{2\} &= \langle v_2^2 + \delta_2 \rangle , \\ c_2\{4\} &= \langle v_2^4 + \delta_4 + 4v_2^2\delta_2 + 2\delta_2^2 \rangle - 2\langle v_2^2 + \delta_2 \rangle^2 , \end{aligned} \tag{2.59}$$

with non-flow contributions δ_2 and δ_4. Here, since for v_n the particle average within one event has been carried out already, the brackets $\langle \cdot \rangle$ denote the average over all events.

As we have seen, in general, the cumulants contain both flow and non-flow contributions. It is shown in [BDO01] that the lowest-order estimates for the Fourier coefficients, $v_n\{2k\}$, can be calculated from the measured cumulants $c_n\{2k\}$ when ignoring all non-flow effects: $v_n\{2\} = \sqrt{c_n\{2\}}$ and $v_n\{4\} = \sqrt[4]{-c_n\{4\}}$. From this the estimates for the elliptic flow are calculated from $c_2\{2k\}$ when setting all non-flow contributions to zero, $\delta_{2k} = 0$. From Eq. (2.59) we find:

$$\begin{aligned} v_2\{2\} &= \sqrt{\langle v_2^2 \rangle} , \\ v_2\{4\} &= \sqrt[4]{2\langle v_2^2 \rangle^2 - \langle v_2^4 \rangle} . \end{aligned} \tag{2.60}$$

The conclusion is that, instead of v_n, actually the all-event average over $\langle v_n^2 \rangle$ is measured. As we will see in the next Section 2.2.2, experimental results for the elliptic flow are usually derived from measuring two or four-particle correlations.

2.2.2. Hydrodynamic description of the quark-gluon plasma

The quark-gluon plasma (QGP) produced at RHIC and LHC can be described as an almost-perfect fluid. Hydrodynamics is commonly used to describe physical systems consisting of $N \approx 10^{23}$ particles. In heavy-ion collisions the typical number of produced particles is only $10^4 - 10^5$ assuming a central collision where the number of participants is $N_{\text{part}} \approx 400$. Nevertheless, one uses relativistic, dissipative hydrodynamics to simulate the (elliptic) flow, cf. for instance [DT08] and references therein. These simulations turn out to work successfully. Comparing the simulated and measured elliptic flow of charged hadrons one can extract some (constant) ratio η/s. This ratio has been found to be small, $\eta/s < 5/4\pi$ [Son13], as we will describe in the following. The energy-momentum tensor of a perfect fluid reads

$$T^{\mu\nu} = (\epsilon + P)u^\mu u^\nu - P g^{\mu\nu} , \tag{2.61}$$

with ϵ being the energy density, P the pressure, and u^μ denoting the *four velocity*

$$u^\mu = \frac{dx^\mu}{d\tau} , \tag{2.62}$$

---

[16]We emphasize that in contrast to the double brackets for $c_n\{2k\}$, for the Fourier coefficients v_n no average over all events is carried out.

2.2. Heavy-ion collisions and the quark-gluon plasma

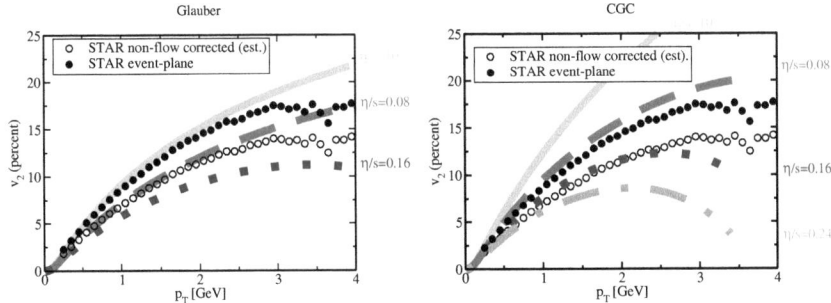

Figure 2.5.: Elliptic flow v_2 as function of transverse momentum from hydrodynamic simulations with different initial conditions from the Glauber model (left) and the color-glass condensate model (right), see the discussion in the text. Both figures have been taken from [LR08].

where τ is the proper time. It is normalized to $u_\mu u^\mu = 1$. The quark-gluon plasma described by hydrodynamics is assumed to be in local equilibrium. All thermodynamic quantities are actually fields, i.e. functions defined on the Minkowski space. Dissipative effects are described by the *dissipative tensor* $\tau^{\mu\nu}$ that extends the energy-momentum tensor to

$$T^{\mu\nu} = (\epsilon + P)u^\mu u^\nu - P g^{\mu\nu} + \tau^{\mu\nu} \,. \tag{2.63}$$

The parameterization of $\tau^{\mu\nu}$ is based on the assumptions that only first-order derivatives of the four velocity are relevant. Using the definitions $\Delta^{\mu\nu} = g^{\mu\nu} - u^\mu u^\nu = \Delta^{\nu\mu}$ and $\partial_\perp^\mu = \Delta^{\mu\nu}\partial_\nu$, one finds the general parameterization [YHM08, Wei72]:

$$\tau^{\mu\nu} = \eta \left[\partial_\perp^\mu u^\nu + \partial_\perp^\nu u^\mu - \frac{2}{3}\Delta^{\mu\nu}(\partial_\perp \cdot u) \right] + \zeta \Delta^{\mu\nu}(\partial_\perp \cdot u) \,, \tag{2.64}$$

where its traceless part is described by the *shear viscosity* η. The non-traceless part is parameterized by the *bulk viscosity* ζ which is assumed to be small compared to the shear viscosity. This is because the bulk viscosity vanishes in conformal theories since it describes dissipative effects arising just from a rescaling of the system. Therefore, in the QGP the bulk viscosity is expected to be much smaller than the shear viscosity since high-T QCD is almost conformal. This argument is also supported by lattice-QCD calculations [Mey08].

Shear and bulk viscosity as introduced in Eq. (2.64) are constant or temperature-dependent

$\eta/s|_{\text{extract}}$	Glauber model	CGC model
non-flow corrected (est.)	0.08	0.16
event-plane	10^{-4}	0.08

Table 2.4.: Resulting constant ratios η/s from Fig. 2.5 using different models for the initial conditions in the hydrodynamic simulation and methods to extract the elliptic-flow coefficient from experimental data. Note that $0.08 = 1/4\pi$ refers to the AdS/CFT benchmark.

2. Quantum Chromodynamics and the Quark-Gluon Plasma

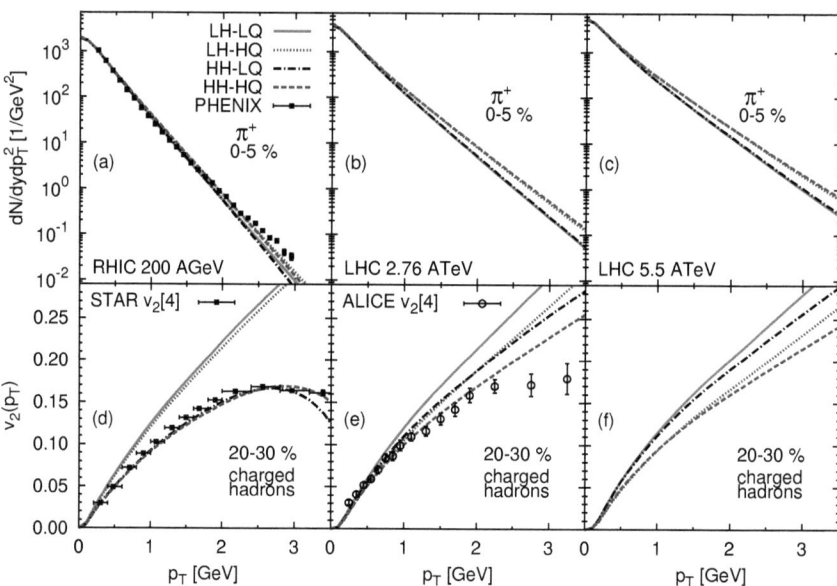

Figure 2.6.: Comparison between experimental results for the elliptic flow $v_2(p_\mathrm{T})$ and hydrodynamic simulations using non-constant parameterizations of $\eta/s(T)$ as shown in Fig. 2.7. See the discussion in the text. Figure taken from [NDH+11]. We focus on the panels (d) to (f) and do not discuss the particle production presented in the figure as well. For panel (f) no LHC data is available so far, the 2015 upgrade will provide $\sqrt{s_\mathrm{NN}} = 5.1$ TeV rather than $\sqrt{s_\mathrm{NN}} = 5.5$ TeV.

parameters when performing hydrodynamic simulations. As we will see in Section 2.3, the Kubo formalism can be used to *calculate* the viscosity coefficients within a quantum field theory. The Kubo formula for the shear viscosity is usually written in terms of the traceless part of the energy-momentum tensor, the so-called *viscous-stress tensor*. It is defined by

$$\pi_{\mu\nu} = \left(\Delta_{\mu\rho} \Delta_{\nu\sigma} - \frac{1}{3} \Delta_{\mu\nu} \Delta_{\rho\sigma} \right) T^{\rho\sigma} , \qquad (2.65)$$

from where it is easily seen that $\pi^\mu_\mu = 0$ as a consequence of $\Delta^\mu_\mu = 3$ and $\Delta_{\mu\alpha} \Delta^\mu_\beta = \Delta_{\alpha\beta}$.

In the Bjorken spacetime picture, relativistic hydrodynamics can be applied for $\tau > \tau_0 \approx 1$ fm, cf. Fig. 2.4. Initial conditions for energy and entropy density, ϵ and s, respectively, need to be calculated externally. There exist two main approaches: the Glauber model [MRSS07] and the color-glass-condensate model [ILM01, FILM02, GIJMV10]. Apart from initial conditions, also the equation of state, i.e. the relationship between energy density and pressure, needs to be specified. In many applications one uses just the extreme case of an ideal gas, $\epsilon = 3P$, neglecting interaction corrections. Incorporating energy-momentum conservation, $\partial_\mu T^{\mu\nu}$, and the second law of thermodynamics, $\partial_\mu(su^\mu) \geq 0$, hydrodynamic simulations can be performed providing direct results for the flow coefficients v_n defined in Eq. (2.51), cf. for instance [LR08, SJG11]. In Fig. 2.5 the comparison between hydrodynamic simulations and elliptic-flow data from the STAR

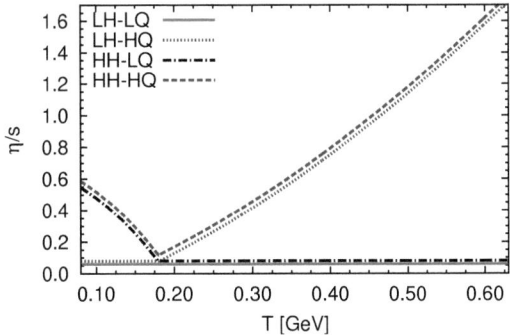

Figure 2.7.: Parameterized temperature dependence of the ratio η/s for hydrodynamical simulations presented in Fig. 2.6. See the discussion in the text. Figure taken from [NDH+11].

experiment are shown. For the left panel initial conditions from the Glauber model have been used, in the right panel the CGC model has been used resulting in different curves for $v_2(p_T)$. In addition, there are experimental uncertainties from the extraction method: the event-plane method systematically results in smaller values for the elliptic flow compared to the estimates where non-flow effects have been corrected. In conclusion, the extraction of a constant ratio η/s suffers from large uncertainties as summarized in Table 2.4. However, the main observation of a small ratio $\eta/s = \mathcal{O}(1/4\pi)$ is valid within these uncertainties.

As we have already discussed, the shear viscosity is not a constant but depends on temperature and density (quark chemical potential). In Fig. 2.6 we show curves from [NDH+11], where results from hydrodynamic simulations for $v_2\{4\}(p_T)$ from four-particle correlations are compared to data from the STAR [Tan08] and ALICE [A+10] experiment in panel (d) and (e), respectively. There are four different parameterizations used for $\eta/s(T)$ shown in Fig. 2.7: a constant and decreasing ratio in the low-T hadronic sector, LH and HH, respectively, combined with a constant and rising ratio in the high-T quark sector, LQ and HQ, respectively. These parameterizations have been motivated by the fact that the ratio $\eta/s(T)$ of a hadron gas decreases with increasing temperature, cf. for instance our calculation of the shear viscosity of a hot pion gas [LKW12], and references therein. In contrast, results from hard thermal loop calculations suggest a rising ratio $\eta/s(T)$ in the high-T region, cf. [AMY00, AMY03] and the discussion of our final results in Section 6.3. From Fig. 2.6(d) it can be seen that only the non-constant parameterizations of η/s describe the STAR data points correctly. Switching between the LQ and HQ parameterization does not affect the hydrodynamic results dramatically. It can be concluded that at RHIC energies, $\sqrt{s_{\text{NN}}} = 200$ GeV, the hydrodynamic simulations are almost insensitive to details of the high-T region. This changes when approaching LHC energies as seen in panels (e) and (f) with $\sqrt{s_{\text{NN}}} = 2.76$ TeV and $\sqrt{s_{\text{NN}}} = 5.5$ TeV, respectively. The ALICE data points indicate that parameterizations with a non-constant ratio η/s in the high-T region is favored. In contrast to panel (d), the sensitivity of the results from hydrodynamic simulations to switching between the LH and HH parameterization is much less pronounced, especially in the low-p_T region. However, the overall conclusion of this discussion is the necessity for a field-theoretical calculation of $\eta/s(T)$ instead of using simple parameterizations. The main goal of this thesis is the derivation of thermal dependences of the ratio η/s calculated within a large-N_c NJL model.

2. Quantum Chromodynamics and the Quark-Gluon Plasma

2.3. Kubo formalism

2.3.1. Transport coefficients from linear-response theory

For physical systems that are not in thermodynamic equilibrium there exist two main microscopic approaches: kinetic theory and Kubo formalism. The first approach is based on the Boltzmann equation which describes the (time) evolution of particle distribution functions. In the literature one usually applies this formalism using certain approximation schemes where the relaxation-time ansatz is the most common one. In Section 4.3 we will briefly introduce its main concepts and discuss how this approximation compares to the Kubo formalism. In this formalism the macroscopic coefficients of dissipative processes can be calculated within a given quantum-field theory from retarded correlators. Later, in Chapter 4, we will evaluate these correlators within the Nambu–Jona-Lasinio (NJL) model investigating their functional and numerical properties. However in this section we first review how the Kubo formalism can be deduced from Zubarev's statistical operator [Zub74, HST84]. Our treatment is based on [Lan10, LKW12] where we have already discussed this formalism in more detail. We start with introducing the four vector

$$F^\mu = \beta_s u^\mu , \qquad (2.66)$$

where β_s denotes the *inverse proper temperature*,

$$\beta_s = \frac{1}{T_s} = \frac{\gamma}{T} , \qquad (2.67)$$

with the Lorentz factor $\gamma = (1 - \boldsymbol{v}^2)^{-1/2}$, and u^μ being the four-velocity (2.62). In the reference frame of the heat bath, $\boldsymbol{v} = 0$, the proper temperature is just the standard temperature. In a comoving frame one has $T_s > T$. We emphasize that, by construction, F^μ transforms under Lorentz transformations indeed as a four-vector. Now we can introduce the statistical operator in Schrödinger picture (i.e. $\dot{\rho} = 0$),

$$\rho(t) = \frac{1}{Q} \exp\left[-A(t) + B(t) \right] , \qquad (2.68)$$

with $Q = \mathrm{Tr}\exp\left[-A + B \right]$ ensuring that the statistical operator is normalized by $\mathrm{Tr}\,\rho(t) = 1$. We have decomposed the operator into the equilibrium part, A, and some part describing the deviation from equilibrium, B:

$$\begin{aligned} A(t) &= \int \mathrm{d}^3\boldsymbol{x}\, F^\mu(t,\boldsymbol{x}) T_{0\mu}(t,\boldsymbol{x}) , \\ B(t) &= \int \mathrm{d}^3\boldsymbol{x} \int_{-\infty}^{t} \mathrm{d}t'\, T_{\mu\nu}(t',\boldsymbol{x}) \partial^\mu F^\nu(t,\boldsymbol{x}) . \end{aligned} \qquad (2.69)$$

We note that both these terms are Lorentz scalars and one realizes that $A(t) = \beta H$ describing the standard equilibrium part of the system, with H denotes its Hamiltonian. In the operator $B(t)$ the tensor $\partial^\mu F^\nu$ describes deviations from thermodynamic equilibrium. We call this tensor *dissipative force* and assume that it is small enough, allowing for an expansion[17] of the statistical

---

[17] For non-commutating operators A and B we use the Baker-Campbell-Hausdorff formula,

$$e^{-A+B} = e^{-A}\left(1 + \int_0^1 \mathrm{d}\xi\, e^{A\xi} B e^{-A\xi} + \mathcal{O}(B^2)\right),$$

and apply it to both the numerator and denominator of ρ.

operator for $\langle A \rangle \gg \langle B \rangle$:

$$\rho = \left(1 + \int_0^1 \mathrm{d}\xi\, \mathrm{e}^{A\xi} B \mathrm{e}^{-A\xi} - \langle B \rangle_0 \right) \rho_0 \,, \tag{2.70}$$

where $\langle \cdot \rangle = \mathrm{Tr}\,(\rho\,\cdot)$ and $\langle \cdot \rangle_0 = \mathrm{Tr}\,(\rho_0\,\cdot)$ denote the thermal expectation value with respect to the full statistical operator ρ and equilibrium statistical operator $\rho_0 = \rho|_{B=0}$, respectively. As the detailed calculations in [HST84] and [Lan10] show, one can extract the Kubo formula for the shear viscosity from the linear response of the energy-momentum tensor to the dissipative force:

$$\eta(\omega; t, \boldsymbol{x}) = \frac{\beta}{10} \int_t^\infty \mathrm{d}t'\, \mathrm{e}^{\mathrm{i}\omega t'} \int \mathrm{d}^3 x'\, \left(\pi_{\mu\nu}(t', \boldsymbol{x}'), \pi^{\mu\nu}(t, \boldsymbol{x})\right), \tag{2.71}$$

where the viscous-stress tensor $\pi_{\mu\nu}$ denotes the traceless part of the energy momentum tensor, defined in Eq. (2.65). We have introduced the correlator $(\cdot, \cdot)$ in the integrand of Eq. (2.71) by

$$(X(t), Y(t')) = \frac{1}{\beta} \int_0^\beta \mathrm{d}\xi\, \langle X(t) [\mathrm{e}^{-\xi H} Y(t') \mathrm{e}^{\xi H} - \langle Y(t') \rangle_0] \rangle_0 \,, \tag{2.72}$$

where its structure is induced by the expanded statistical operator in Eq. (2.70). We note that due to the conjugation operation known from the Heisenberg picture of operators one has in the Matsubara formalism with imaginary time:

$$\mathrm{e}^{-\beta H} X(t) \mathrm{e}^{\beta H} = X(t + \mathrm{i}\beta) \,. \tag{2.73}$$

This implies $\langle X(t) Y(t' + \mathrm{i}\beta) \rangle_0 = \langle Y(t') X(t) \rangle_0$ and shows that this correlator is symmetric in its arguments:

$$(X(t), Y(t')) = (Y(t'), X(t)) \,. \tag{2.74}$$

One can prove this identity by a small and straightforward calculation.

2.3.2. Ladder-diagram resummation in the Kubo formalism

The Kubo formula (2.71) for the shear viscosity results from linear-response theory and needs to be evaluated for a given Lagrangian. We have done this in a toy model ($\lambda \phi^4$ theory) and for a hot pion gas within the framework of chiral perturbation theory [Lan10, LKW12]. The so-called *skeleton expansion* was used, i.e. we have expanded the four-point correlator $\eta(\omega; t, \boldsymbol{x})$ in Feynman diagrams including *full propagators* only. It has been known that

> "... diagrammatic evaluation of transport coefficients is a remarkably inefficient approach. An infinite set of rather complicated diagrams must be summed, merely to obtain the leading weak coupling behavior." [JY96]

In this section we report how the divergent infrared behavior of the shear viscosity leads to the necessity of ladder-diagram resummation, even when dealing with a simple toy model like $\lambda \phi^4$ theory. This formalism is described in great detail in [Jeo95] and its connection to the kinetic approach for evaluating transport coefficients is investigated in [JY96]. Already at this stage we emphasize the main point of this analysis relevant for our purposes: ladder-diagram resummation is necessary when evaluating the shear viscosity in some weak-coupling limit. In the case of the NJL model (cf. Section 4.1) the corresponding ladder-diagrams are subleading in a large-N_c expansion. In addition, an explicit numerical check of the non-perturbative nature of the NJL model is performed in Section 4.2.4. Compare this also to the discussion in Section 4.3.

2. Quantum Chromodynamics and the Quark-Gluon Plasma

(a) One-loop diagram   (b) One-rung ladder diagram

Figure 2.8.: Skeleton diagrams for the diagrammatic evaluation of shear viscosity in a weakly coupled $\lambda\phi^4$ theory. The dashed double lines denote fully dressed Bose propagators.

We follow now the arguments of Jeon and Yaffe and consider the toy model

$$\mathcal{L} = \frac{1}{2}\left(\partial_\mu \phi\right)\left(\partial^\mu \phi\right) - \frac{1}{2}m_0^2 \phi^2 - \lambda \phi^4 \ . \tag{2.75}$$

Assuming a weak coupling $\lambda \ll 1$, the naive expectation for the leading-order contribution to the shear viscosity would be described by a one-loop skeleton diagram shown in Fig. 2.8(a) [HST84, Jeo93]. Doing so, one finds

$$\eta \sim \frac{1}{\Gamma} \sim \frac{1}{\lambda^2} \ , \tag{2.76}$$

where Γ denotes the spectral width of the fully dressed boson. This shows that η becomes large when the coupling parameter is small, leading to a divergent viscosity in the limit of a free quantum gas. There, the mean free time, $\tau \sim 1/\Gamma$, diverges. The divergence originates from *pinched poles* arising from the fully dressed boson propagator, $p_0 = \pm E \pm i\Gamma$, where $E = \sqrt{p^2 + m^2}$ denotes the on-shell energy of the dressed boson. There are always two poles separated from each other by $2i\Gamma$ in the positive and negative imaginary half plane, respectively. It is convenient to express the (dressed) propagators in spectral representation[18], so it becomes apparent from the frequency integral[19] that in the weak-coupling limit this pole structure leads to the scaling $\eta \sim 1/\Gamma$.

Still following the arguments by Jeon and Yaffe we now consider also ladder diagrams with n additional rungs compared to the one-loop diagram, cf. Fig 2.8(b) where the case $n = 1$ is shown. Naively, this skeleton diagram is suppressed because of the appearance of two additional coupling constants λ^2. In fact, when evaluating the shear viscosity, things are more involved. It turns out that only the imaginary part of the skeleton diagrams actually contributes to the shear viscosity which means that one has to *cut the diagrams*. For the ladder-diagram of order n this means that there are n factors of the mean-free time $\tau \sim 1/\Gamma$, because such a cut diagram can be interpreted as a n-particle exchange $2 \to 2$ scattering process [Jeo95]. At leading order one has $\Gamma \sim \lambda^2$, and one can conclude that all ladder diagrams contribute at the same order as the one-loop skeleton diagram shown in Fig 2.8(a):

$$\left(\lambda^2\right)^n \cdot \tau^n \sim \lambda^{2n} \cdot \frac{1}{\Gamma^n} \sim 1 \ . \tag{2.77}$$

As a consequence, an infinite set of ladder diagrams must be summed in order to arrive at the full leading-order result of the weakly-coupled toy model. Without going into technical details

---
[18] Details of the spectral representation are discussed in the Appendix A.3.
[19] In practice, this integration is carried out using residual calculus, so the integral transforms to a sum over residues which is then expanded in the weak-coupling limit.

of how such a resummation can be performed, we want to state Jeon's' final result [Jeo95]:

$$\eta = a \frac{T^3}{\lambda^2} \left[1 + \mathcal{O}(\sqrt{\lambda}) + \mathcal{O}(\frac{m}{T}) \right], \tag{2.78}$$

with $a_{\text{resum}} = 3040$ a purely numerical dimensionless number. Ignoring the resummation does not change the functional shape of $\eta(T)$, but its coefficient is underestimated by roughly a factor of four: $a_{1-\text{loop}} = 733$.

As also highlighted by Jeon and Yaffe, the importance of the resummed ladder diagrams becomes even more evident when realizing that the resummed expression for the shear viscosity just coincides with the result from kinetic theory. We will discuss this approach using the Boltzmann equation later in more detail in Section 4.3. The central assumption when applying kinetic theory for the description of (relativistic) fluids is that the collision time is much smaller than to the mean free time between two collisions. This condition is fulfilled in a weakly coupled theory. The non-perturbative NJL model does not meet this condition, cf. Section 4.2.4. This makes us omitting ladder-diagram resummation in our calculations.

We conclude this section with giving a brief analogy between shear viscosity and the electrical resistance of a circuit. Consider a physical system where several dissipative processes operate, e.g. by having different coupling constants or different types of Yukawa interactions. Each process induces a spectral width, Γ_i, which defines a corresponding shear viscosity $\eta_i \sim 1/\Gamma_i$, where we work again in the weak-coupling limit. The full spectral width $\Gamma = \Gamma_1 + \Gamma_2 + \ldots$ can be written as simple sum of the individual widths, assuming the processes to be independent. From this, the full shear viscosity which is related to the entire fluid calculates as

$$\eta = \left(\eta_1^{-1} + \eta_2^{-1} + \ldots \right)^{-1}. \tag{2.79}$$

From this one can see that the shear viscosity behaves as the resistances in a parallel circuit and the individual spectral widths can be interpreted as electric current. With N equal but independent dissipative processes the resulting shear viscosity becomes $\eta = \eta_0/N$, i.e. it decreases with increasing multiplicity. This scaling property is useful for constructing field-theoretical models which violate the AdS/CFT benchmark as discussed previously in Section 2.1.3.

3. The Nambu–Jona-Lasinio model

> *"Nowadays, the principle of spontaneous symmetry breaking is the key concept in understanding why the world is so complex as it is, in spite of the many symmetry properties in the basic laws that are supposed to govern it. The basic laws are very simple, yet this world is not boring; that is, I think, an ideal combination."[Nam08]*
>
> Yoichiro Nambu, Nobel Presentation Ceremony 2008

Originally, the Nambu–Jona-Lasinio (NJL) model has been introduced in 1961 by Y. Nambu and G. Jona-Lasinio in order to explain the nucleon mass [NJL61a, NJL61b]. The title "Dynamical model of elementary particles based on an analogy with superconductivity" already displays a mechanism for mass generation: in analogy to the Bardeen-Cooper-Schrieffer (BCS) theory of superconductivity, non-perturbative self-energy processes lead to a finite mass described by so-called gap equations. Nambu was awarded the Nobel Prize 2008 *"for the discovery of the mechanism of spontaneous broken symmetry in subatomic physics"*. The former and recent success of the NJL model is based on the appearance of spontaneous chiral symmetry breaking in this model.

Nowadays, QCD as part of the Standard Model is accepted as the fundamental theory of the strong interaction. In this context the NJL model is reinterpreted as a dynamical model of quarks and their composites: mesons [KLVW90a, KLVW90b] and baryons [IBY93, Ish98]. For reviews of the NJL model we refer to the literature [VW91, Kle92, HK94, Bub05]. The NJL model is a purely fermionic theory where all gluonic degrees of freedom have been integrated out. Nevertheless, as we will discuss in detail and employ intensely, the color symmetry of QCD affects the structure of the NJL model by scaling properties of the coupling constants. In fact the original local color gauge symmetry of QCD is replaced by a global color symmetry in the NJL model.

3.1. General N_f, N_c-NJL Lagrangian in the chiral limit

To construct the NJL Lagrangian as a model for QCD we start with its fundamental color currents $J_\mu^a = \bar{\psi}\gamma_\mu T^a \psi$, where T^a ($a = 1, \ldots, N_c^2 - 1$) are the generators[20] of $SU(N_c)$. In the NJL model the gluonic degrees of freedom are integrated out. The interaction between quarks becomes local:

$$\mathcal{L}_{cc} = -G_c(\bar{\psi}\gamma_\mu T^a \psi)^2 \,, \qquad (3.1)$$

with some dimensionful, effective coupling strength G_c containing all the gluon dynamics. The first step towards the NJL Lagrangian is a general channel analysis of color-color currents. As usual we assume the quarks to be realized in the fundamental representation F. There are two different product representations:

$$F \otimes \bar{F} = \mathbb{1} \oplus A \,, \quad \text{and} \quad F \otimes F = r_s \oplus \bar{r}_a \,, \qquad (3.2)$$

---

[20]In Appendix A.1 we briefly review the most important properties of the Lie group $SU(N)$.

3. The Nambu–Jona-Lasinio model

where the physical case reads

$$3 \otimes \bar{3} = 1 \oplus 8, \quad \text{and} \quad 3 \otimes 3 = 6_s \oplus \bar{3}_a. \tag{3.3}$$

It is well-known that the gluon-exchange interaction in the color-singlet channel is attractive, whereas it is repulsive in the octet channel. In the following we discuss general N_c-structure of the relevant interactions. The quark-antiquark current decomposes as

$$F \otimes \bar{F} = \Box \otimes \{\text{tableau}\} (N_c-1) = \cdot \oplus (N_c-1) \{\text{tableau}\}, \tag{3.4}$$

with dimensions $\dim(1) = 1$ and $\dim(A) = N_c^2 - 1$. From the Young tableau of the adjoint representation it follows that this representation is always real: $\bar{A} = A$. The decomposition of the quark-quark current reads

$$F \otimes F = \Box \otimes \Box = \Box\Box \oplus \overset{\Box}{\Box}, \tag{3.5}$$

with dimensions $\dim(r_s) = \frac{N_c(N_c+1)}{2}$ and $\dim(r_a) = \frac{N_c(N_c-1)}{2}$. Directly from the Young tableaux we find criteria for Casimir operators to be fulfilled in our general treatment:

1. The totally antisymmetric representation is the anti-fundamental one for the physical case $N_c = 3$, i.e. $\bar{r}_a|_{N_c=3} = \bar{F}$.

2. The totally antisymmetric representation is the trivial one for $N_c = 2$, i.e. $\bar{r}_a|_{N_c=2} = 1$.

3. The totally symmetric representation is the adjoint one for $N_c = 2$, i.e. $r_s|_{N_c=2} = A$.

These properties provide a consistency check for the representation-dependent Casimir operators $C_2(r)$ summarized in Table 3.1.

Apart from the two SU(N_c) representations in Eq. (3.2) there is a third one, generated by two antiquarks: $\bar{F} \otimes \bar{F}$. We expect for its decomposition in irreducible terms just $\bar{r}_s \oplus r_a$. Indeed, we find in terms of Young tableaux:

$$\bar{F} \otimes \bar{F} = \{\text{tableau}\} (N_c - 1) \oplus \{\text{tableau}\} (N_c - 2) = r_a \oplus \bar{r}_s. \tag{3.6}$$

We also verify their dimensions using the so-called *hook-length formula*:[21]

$$\dim\left(\{\text{tableau}\} (N_c - 1)\right) = \frac{N_c! \frac{(N_c+1)!}{2}}{(N_c - 1)! N_c!} = \frac{N_c(N_c + 1)}{2} = \dim(r_s), \tag{3.7}$$

---

[21] This and more details about Young tableaux can be found in standard textbooks, e.g. [FH91].

3.1. General N_f, N_c-NJL Lagrangian in the chiral limit

r	$\mathbb{1}$	F	$\bar{\mathrm{F}}$	$\mathrm{r_s}$	$\bar{\mathrm{r}}_\mathrm{a}$	A
Young tableau	·	□	$N-1\left\{\begin{array}{c}\square\\\square\\\square\end{array}\right.$	□□	$\begin{array}{c}\square\\\square\end{array}$	$N-1\left\{\begin{array}{c}\square\square\\\square\\\square\end{array}\right.$
$n = \dim(\mathrm{r})$	1	N	N	$\frac{N(N+1)}{2}$	$\frac{N(N-1)}{2}$	N^2-1
$C(\mathrm{r})$	0	$\frac{1}{2}$	$\frac{1}{2}$	$\frac{N+2}{2}$	$\frac{N-2}{2}$	N
$C_2(\mathrm{r})$	0	$\frac{N^2-1}{2N}$	$\frac{N^2-1}{2N}$	$\frac{(N+2)(N-1)}{N}$	$\frac{(N+1)(N-2)}{N}$	N
$C_2(\mathrm{r})\|_{N=2}$	0	$\frac{3}{4}$	$\frac{3}{4}$	2	0	2
$C_2(\mathrm{r})\|_{N=3}$	0	$\frac{4}{3}$	$\frac{4}{3}$	$\frac{10}{3}$	$\frac{4}{3}$	3

Table 3.1.: Relevant irreducible representations of SU(N) with their dimensions and Casimir operators. For the values of $C_2(\mathrm{r_s})$ and $C_2(\bar{\mathrm{r}}_\mathrm{a})$ discussed in the text we refer to [Kob73], the rest can be found or easily calculated from standard textbooks.

and also for $\bar{\mathrm{r}}_\mathrm{s}$:

$$\dim\left(\begin{array}{c}\square\\\square\end{array}(N_\mathrm{c}-2)\right) = \frac{\frac{N_\mathrm{c}!}{2}}{(N_\mathrm{c}-2)!} = \frac{N_\mathrm{c}(N_\mathrm{c}-1)}{2} = \dim(\bar{\mathrm{r}}_\mathrm{a}) \ . \tag{3.8}$$

In order to derive the general matrix elements of color-color currents we need to prepare all Casimir operators. $C_2(\mathbb{1})$, $C_2(\mathrm{F})$, $C_2(\bar{\mathrm{F}})$ and $C_2(\mathrm{A})$ can be found in standard textbooks, e.g. [Gre05, FH91]. For the Casimirs of the totally (anti)symmetric representations we start with a more general result from [Kob73] where the Casimir operators of order p of U(N_c) and SU(N_c) groups are discussed. It is found that

$$\bar{C}_\alpha^{(p)}(h) = h(N_\mathrm{c}-\alpha)(N_\mathrm{c}+\alpha h)\frac{(N_\mathrm{c}-\alpha)^{p-1}(N_\mathrm{c}+\alpha h)^{p-1}-(-h)^{p-1}}{N_\mathrm{c}^p(N_\mathrm{c}+\alpha h-\alpha)} \ , \tag{3.9}$$

where $\bar{C}$ displays that this result refers to the SU(N_c) case. The symmetric and antisymmetric representations are denoted by $\alpha = +1$ and $\alpha = -1$, respectively. For our purpose we need the second-order Casimir, $p=2$, with $2C_2 = \bar{C}_\alpha^{(2)}$. For $h=1$ we find

$$\bar{C}_\alpha^{(2)}(1) = (N_\mathrm{c}-\alpha)(N_\mathrm{c}+\alpha)\frac{(N_\mathrm{c}-\alpha)(N_\mathrm{c}+\alpha)+1}{N_\mathrm{c}^3} = \frac{N_\mathrm{c}-1}{N_\mathrm{c}} \ , \tag{3.10}$$

which is indeed independent of $\alpha \in \{-1,1\}$. We identify here twice the quadratic Casimir of the fundamental and anti-fundamental representation. It remains true for a general p that the Casimir evaluated at $h=1$ is independent of α. For $h=2$ the expressions for the symmetric

3. The Nambu–Jona-Lasinio model

and antisymmetric representation are found:

$$\bar{C}_1^{(2)}(2) = \frac{2(N_c + 2)(N_c - 1)}{N_c} = 2C_2(r_s),$$
$$\bar{C}_{-1}^{(2)}(2) = \frac{2(N_c + 1)(N_c - 2)}{N_c} = 2C_2(\bar{r}_a).$$
(3.11)

Let now $C_2(1,2)$ denote the quadratic Casimir operator of an irreducible component in a product representation $r_1 \otimes r_2$ with Casimirs $C_2(1)$ and $C_2(2)$, respectively. With $n = \dim(r)$ and

$$T_a(i)T_a(i) = C_2(i)\, \mathbb{1}_{n \times n}, \quad \big(T_a(i) + T_a(j)\big)^2 = C_2(i,j)\, \mathbb{1}_{n \times n},$$
(3.12)

we find

$$\langle T_a(1) \cdot T_a(2) \rangle = \frac{1}{2}\big(C_2(1,2) - C_2(1) - C_2(2)\big).$$
(3.13)

Using the Casimirs listed in Table 3.1 the matrix elements for the color-color currents are summarized in Table 3.2 for an arbitrary number of colors $N_c \geq 1$. As an example we derive for the color-singlet case:

$$C_2(1) = C_2(2) = C_2(F) = \frac{N_c^2 - 1}{2N_c},$$
(3.14)

and $C_2(1,2) = C_2(\mathbb{1}) = 0$. We therefore find

$$\langle \mathbb{1} \rangle = \langle T_a(1) \cdot T_a(2) \rangle = \frac{N_c^2 - 1}{2N_c} = \frac{1}{2}\left(\frac{1}{N_c} - N_c\right),$$
(3.15)

which is half the matrix element given in Table 3.2. We conclude from this table that in the large-N_c limit the attractive color-singlet channel, $\mathbb{1}$, becomes dominant, whereas the repulsive adjoint channel, A, becomes negligible. The $\bar{r}_a$ channel is attractive as well, but bounded in its strength. Only the r_s channel stays repulsive for all $N_c \geq 1$. In the whole thesis we will therefore consider, on the level of the Lagrangian, quark-antiquark currents in the color-singlet channel only. This approach becomes exact if $N_c \to \infty$.

The next steps for constructing the NJL Lagrangian are based on the set $\mathcal{A}$ of Noether currents induced by the (classical) symmetry pattern of QCD in its chiral limit:

$$\mathcal{C} \times \mathcal{P} \times \mathcal{T} \times \mathrm{SU}(N_c) \times \mathrm{SU}(N_f)_L \times \mathrm{SU}(N_f)_R \times \mathrm{U}(1)_V \times \mathrm{U}(1)_A.$$
(3.16)

Besides the discrete $\mathcal{C} \times \mathcal{P} \times \mathcal{T}$ symmetries there remains a $\mathrm{SU}(N_c)$ color symmetry which becomes *global* after integrating out all gluonic degrees of freedom resulting in effective four-

channel	$\mathbb{1}$	A	r_s	$\bar{r}_a$
matrix element	$\frac{1}{N_c} - N_c$	$\frac{1}{N_c}$	$1 - \frac{1}{N_c}$	$-\left(1 + \frac{1}{N_c}\right)$
range for $N_c \geq 1$	$(-\infty, 0]$	$(0, 1]$	$[0, 1)$	$[-2, -1)$
physical value $N_c = 3$	$-\frac{8}{3}$	$\frac{1}{3}$	$\frac{2}{3}$	$-\frac{4}{3}$
large-N_c limit	$-N_c$	0	1	-1

Table 3.2.: Matrix elements for $\mathrm{SU}(N_c)$ color-color currents in their four different channels

3.1. General N_f, N_c-NJL Lagrangian in the chiral limit

fermion vertices. The most general four-fermion vertex which is the square of some bilinear $\bar{\psi}\Gamma\psi$. It can be built from six different combinations of vector and axialvector currents [BJM87, BJM88, KLVW90a]:

$$\mathcal{A} = \{V^2 \pm A^2, V_f^2 + A_f^2, V_c^2 \pm A_c^2, V_{cf}^2 + A_{cf}^2\}, \qquad (3.17)$$

where the indices refer to flavor and color space, e.g. $A_{cf}^2 = (\bar{\psi}\lambda_j^c\lambda_i^f\gamma_\mu\gamma_5\psi)^2$. These local NJL interactions do not distinguish any more between direct and exchange terms present in QCD. In order to model both types of interaction terms one applies the Fierz transformation to this effective vertex as we discuss in the following.

Let $\Gamma \in$ Dirac $\otimes$ Flavor $\otimes$ Color describe a general QCD current, e.g. the current A_{cf} from Eq. (3.17) corresponds to $\Gamma = \lambda_j^c \lambda_i^f \gamma_\mu \gamma_5$. Then the Fierz transformation $\mathcal{F}$ of the (direct) vertex $(\bar{\psi}\Gamma\psi)^2$ is defined as crossing operation that exchanges the outgoing quarks:

$$\begin{aligned}\mathcal{F}\left(\bar{\psi}_b\psi_a\bar{\psi}_d\psi_c\sum_k \Gamma_{ba}^k\Gamma_{dc}^k\right) &= -\bar{\psi}_b\psi_a\bar{\psi}_d\psi_c\sum_k \Gamma_{bc}^k\Gamma_{da}^k = \\ &= -\bar{\psi}_b\psi_a\bar{\psi}_d\psi_c\sum_{k,m} c_m^k\Gamma_{ba}^m\Gamma_{dc}^m .\end{aligned} \qquad (3.18)$$

In the second equation we just have rewritten the vertex in the original basis introducing the Fierz coefficients c_m^k. For our purpose we need these coefficients in Dirac space (in particular in four dimensions) and in SU(N) related to flavor and color space. Assuming a basis $\{\Gamma^k\}_{k=1,\dots,N^2}$ with the normalization $\text{tr}(\Gamma^k\Gamma^l) = N\delta^{kl}$, multiplication of the last equation in (3.18) with $\Gamma_{cd}^l\Gamma_{ab}^l$ leads to

$$c_l^k = \frac{1}{N^2}\text{tr}\left(\Gamma^k\Gamma^l\Gamma^k\Gamma^l\right). \qquad (3.19)$$

From this it is evident that the transformation matrix is symmetric: $c_l^k = c_k^l$. Furthermore, by its definition the Fierz transformation is an involution, i.e. $\mathcal{F}^2 = $ id. Usually one defines the Fierz-invariant vertex as

$$^\mathcal{F}\mathcal{L}_4 = \frac{1}{2}(\mathcal{L}_4 + \mathcal{F}(\mathcal{L}_4)), \qquad (3.20)$$

which ensures that an already Fierz-invariant Lagrangian is not modified when applying a Fierz transformation $\mathcal{F}$.

We need to discuss in more detail the four-dimensional Dirac space: A basis $\{\Gamma^k\}_{k=1,\dots,16}$ of Dirac structures satisfying $\text{tr}\left(\Gamma^k\Gamma^l\right) = 4\delta^{kl}$ is given by

$$\{\Gamma^k\}_k = \{\mathbb{1}, \gamma_5, \gamma_0, i\gamma_i, i\gamma_0\gamma_5, \gamma_i\gamma_5, i\sigma^{0i}, \sigma^{ij}\}, \qquad (3.21)$$

where $\sigma^{\mu\nu}$ contains six tensor terms and is defined as usual $\sigma^{\mu\nu} = \frac{i}{2}[\gamma^\mu, \gamma^\nu]$ [PS95]. The Fierz coefficients in Dirac space can be summarized in

$$\begin{pmatrix} (\mathbb{1})_{ij}(\mathbb{1})_{kl} \\ (i\gamma_5)_{ij}(i\gamma_5)_{kl} \\ (\gamma^\mu)_{ij}(\gamma_\mu)_{kl} \\ (\gamma^\mu\gamma_5)_{ij}(\gamma_\mu\gamma_5)_{kl} \\ (\sigma^{\mu\nu})_{ij}(\sigma_{\mu\nu})_{kl} \end{pmatrix} = \begin{pmatrix} 1/4 & -1/4 & 1/4 & -1/4 & 1/8 \\ -1/4 & 1/4 & 1/4 & -1/4 & -1/8 \\ 1 & 1 & -1/2 & -1/2 & 0 \\ -1 & -1 & -1/2 & -1/2 & 0 \\ 3 & -3 & 0 & 0 & -1/2 \end{pmatrix} \begin{pmatrix} (\mathbb{1})_{il}(\mathbb{1})_{kj} \\ (i\gamma_5)_{il}(i\gamma_5)_{kj} \\ (\gamma^\mu)_{il}(\gamma_\mu)_{kj} \\ (\gamma^\mu\gamma_5)_{il}(\gamma_\mu\gamma_5)_{kj} \\ (\sigma^{\mu\nu})_{il}(\sigma_{\mu\nu})_{kj} \end{pmatrix},$$

$$(3.22)$$

3. The Nambu–Jona-Lasinio model

where vector, axialvector and tensor contributions have been rearranged for convenience[22]. We proceed with discussing in more detail the Fierz transformation in SU(N) space also: Let λ_a, $a = 1, \ldots, N^2 - 1$, denote the infinitesimal generators of SU(N) in its fundamental representation. With the definition $\tilde{T}_a = \sqrt{N/2}\,\lambda_a$ the set $\{\Gamma^k\}_k = \{\mathbf{1}, \tilde{T}_a\}$ is a basis of SU(N) satisfying $\mathrm{tr}(\Gamma^k \Gamma^l) = N\delta^{kl}$, cf. Table (A.1) in the Appendix. In this basis the Fierz coefficients read

$$\begin{pmatrix} (\mathbf{1})_{ij}(\mathbf{1})_{kl} \\ (\tilde{T}_a)_{ij}(\tilde{T}_a)_{kl} \end{pmatrix} = \begin{pmatrix} 1/N & 1/N \\ (N^2-1)/N & -1/N \end{pmatrix} \begin{pmatrix} (\mathbf{1})_{il}(\mathbf{1})_{kj} \\ (\tilde{T}_a)_{il}(\tilde{T}_a)_{kj} \end{pmatrix}. \qquad (3.23)$$

As in Dirac space also for SU(N) the determinant of the Fierz transformation is -1, independent of N. Using the generators λ_a instead of $\tilde{T}_a$ yields the commonly used Fierz coefficients:

$$\begin{pmatrix} (\mathbf{1})_{ij}(\mathbf{1})_{kl} \\ (\lambda_a)_{ij}(\lambda_a)_{kl} \end{pmatrix} = \begin{pmatrix} 1/N & 1/2 \\ 2(N^2-1)/N^2 & -1/N \end{pmatrix} \begin{pmatrix} (\mathbf{1})_{il}(\mathbf{1})_{kj} \\ (\lambda_a)_{il}(\lambda_a)_{kj} \end{pmatrix}. \qquad (3.24)$$

We want to remark the following: The set $\mathcal{A}$ of currents in Eq. (3.17) has been found restricting the four-fermion vertex to be the *square* of some bilinear. Mathematically we could also allow for a four vertex which is just the product of two bilinears, $\bar{\psi}\Gamma^k\psi$ and $\bar{\psi}\Gamma^l\psi$. This most general form violates Lorentz invariance and is therefore not suited to describe physics in terms of a relativistic quantum field theory. However, in this general case the Fierz crossing operation is described by the coefficients c^{kl}_{mn}:

$$\mathcal{F}\left(\bar{\psi}_b\psi_a\bar{\psi}_d\psi_c\sum_{k,l}\Gamma^k_{ba}\Gamma^l_{dc}\right) = -\bar{\psi}_b\psi_a\bar{\psi}_d\psi_c\sum_{k}\Gamma^k_{bc}\Gamma^l_{da} = -\psi_b\psi_a\psi_d\psi_c\sum_{k,l,m,n}c^{kl}_{mn}\Gamma^m_{ba}\Gamma^n_{dc}. \qquad (3.25)$$

For a basis $\{\Gamma^k\}_{k=1,\ldots,N^2}$, again with the normalization $\mathrm{tr}(\Gamma^k\Gamma^l) = N\delta^{kl}$, these coefficients read

$$c^{kl}_{mn} = \frac{1}{N^2}\mathrm{tr}(\Gamma^m\Gamma^k\Gamma^n\Gamma^l). \qquad (3.26)$$

For $l = k$ and $n = m$ they reduce to the physical Fierz coefficient c^k_m used in the NJL model.

---

[22] Note that we have shortened the resulting 16×16 matrix (with determinant -1) by summing over the Dirac indices, e.g. $(\gamma^\mu)_{ij}(\gamma_\mu)_{kl}$. Actually, this entry is again a vector:

$$\begin{pmatrix} (\gamma^0)(\gamma_0) \\ (\gamma^1)(\gamma_1) \\ (\gamma^2)(\gamma_2) \\ (\gamma^3)(\gamma_3) \end{pmatrix}_{(ij)(kl)} = \frac{1}{4}\mathbf{1}_{4\times 4}(\mathbf{1})_{il}(\mathbf{1})_{kj} + \frac{1}{4}\mathbf{1}_{4\times 4}(i\gamma_5)_{il}(i\gamma_5)_{kj} +$$

$$+ \frac{1}{4}\begin{pmatrix} 1 & -1 & -1 & -1 \\ -1 & 1 & -1 & -1 \\ -1 & -1 & 1 & -1 \\ -1 & -1 & -1 & 1 \end{pmatrix}\begin{pmatrix} (\gamma^0)(\gamma_0) + (\gamma^0\gamma_5)(\gamma_0\gamma_5) \\ (\gamma^1)(\gamma_1) + (\gamma^1\gamma_5)(\gamma_1\gamma_5) \\ (\gamma^2)(\gamma_2) + (\gamma^2\gamma_5)(\gamma_2\gamma_5) \\ (\gamma^3)(\gamma_3) + (\gamma^3\gamma_5)(\gamma_3\gamma_5) \end{pmatrix}_{(il)(kj)} +$$

$$+ \begin{pmatrix} -1 & -1 & -1 & 1 & 1 & 1 \\ -1 & 1 & 1 & -1 & -1 & 1 \\ 1 & -1 & 1 & -1 & 1 & -1 \\ 1 & 1 & -1 & 1 & -1 & -1 \end{pmatrix}\begin{pmatrix} (\sigma^{01})(\sigma_{01}) \\ (\sigma^{02})(\sigma_{02}) \\ (\sigma^{03})(\sigma_{03}) \\ (\sigma^{12})(\sigma_{12}) \\ (\sigma^{13})(\sigma_{13}) \\ (\sigma^{23})(\sigma_{23}) \end{pmatrix}_{(il)(kj)}.$$

The sum over these vector components gives rise to the coefficient line $(1, 1, -1/2, -1/2, 0)$ in Eq. (3.22).

3.1. General N_f, N_c-NJL Lagrangian in the chiral limit

Applying the Fierz transformation to the currents $\mathcal{A}$ in Eq. (3.17) lead to the following four-fermion vertex, where due to the previous large-N_c discussion only the color-singlet channel is considered explicitly:

$$^{\mathcal{F}}\mathcal{L}_4 = \frac{G_1}{4}\sum_{i=0}^{N_f^2-1}\left[(\bar{\psi}\lambda_i\psi)^2 + (\bar{\psi}\lambda_i i\gamma_5\psi)^2\right] - \frac{G_2}{4}\sum_{i=0}^{N_f^2-1}\left[(\bar{\psi}\lambda_i\gamma_\mu\psi)^2 + (\bar{\psi}\lambda_i\gamma_\mu\gamma_5\psi)^2\right] \\ - \frac{G_3}{4}(\bar{\psi}\lambda_0\gamma_\mu\psi)^2 - \frac{G_4}{4}(\bar{\psi}\lambda_0\gamma_\mu\gamma_5\psi)^2 + \mathcal{L}^{\text{color adjoint}} \ . \quad (3.27)$$

The Fierz transformation in Dirac space (3.22) induces the relative sign between the G_1 and G_2 terms. The previously used symmetry arguments imply that these two coupling constants are not independent: $G_1 = 2G_2$. Since the chiral symmetry holds approximatively only, this restriction is eased resulting in two independent couplings whose ratio G_1/G_2 is close to 2. The Fierz transformation in flavor space gives rise to flavor-singlet vector and axialvector terms which are separated with independent coupling G_3 and G_4. In contrast, corresponding terms for the scalar and pseudoscalar channel do not feature the designated symmetry separately, hence these terms are already covered by the G_1 term.

Inspecting the four-fermion vertex $\mathcal{L}_4$ one observes an unwanted $U(1)_A$ symmetry. On the level of QCD this symmetry is broken by quantum effects (the axial anomaly). Following the general arguments [KKM71, tH76], there is a unique structure which removes the axial symmetry but preserves the chiral symmetry, namely a totally antisymmetric flavor term:

$$\mathcal{L}_{2N_f} \sim \det_{ij}\left[\bar{\psi}_i(1+\gamma_5)\psi_j\right] + \det_{ij}\left[\bar{\psi}_i(1-\gamma_5)\psi_j\right], \quad (3.28)$$

where the determinant refers to flavor space only. For N_f flavors this generates a $2N_f$-vertex which has $N_f! - 1$ possible crossing terms for the Fierz transformation, cf. Fig. 3.1. By construction the 't Hooft interaction is antisymmetric in flavor space. Hence the Fierz transformation has to be performed only in Dirac and color space. The generalized Fierz transformation for an fermionic $2N_f$-vertex is straightforward:

$$\mathcal{F}\left(\prod_{i=1}^{N_f}\bar{\psi}_{a_i}\psi_{b_i}\Gamma^k_{a_ib_i}\right) = \sum_{\pi\in S(N_f)/\{\text{id}\}}\text{sgn}(\pi)\prod_{i=1}^{N_f}\bar{\psi}_{a_i}\psi_{b_{\pi(i)}}\Gamma^k_{a_ib_{\pi(i)}} \ . \quad (3.29)$$

From this the Fierz-invariant vertex is defined as

$$^{\mathcal{F}}\mathcal{L}_{2N_f} = \frac{1}{N_f!}\left(\mathcal{L}_{2N_f} + \mathcal{F}(\mathcal{L}_{2N_f})\right). \quad (3.30)$$

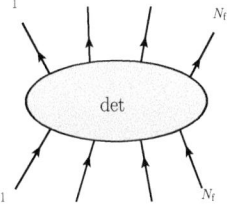

Figure 3.1.: 't Hooft determinant as effective $2N_f$-vertex

3. The Nambu–Jona-Lasinio model

We briefly summarize our general discussion: based on the (approximate) symmetry patterns of QCD the NJL model introduces first the most general four-fermion interaction (square of bilinears) which is consistent with these symmetries. In this vertex all gluonic degrees of freedom have been integrated out. Therefore one applies the Fierz transformation (in Dirac, flavor and color space) in order to include both, direct and exchange terms. Inspecting the N_c-dependences of matrix elements in different channels, the color-singlet channel turns out to be the dominant one in the large-N_c limit. The so constructed four-vertex features an unwanted axial symmetry in the flavor sector, hence the 't Hooft determinant, a $2N_f$-vertex, is introduced in order to remove this symmetry. In total we have:

$$\mathcal{L}_{\text{NJL}} = \mathcal{L}_{\text{kin}} + {}^{\mathcal{F}}\mathcal{L}_{\text{int}} = \mathcal{L}_{\text{kin}} + {}^{\mathcal{F}}\mathcal{L}_4 + {}^{\mathcal{F}}\mathcal{L}_{2N_f} . \tag{3.31}$$

For $N_f = 2$ the 't Hooft interaction is also a four-vertex. In this case, the NJL Lagrangian can be rearranged to recover the original historical Lagrangian shown in Eq. (3.41).

We complete this section with a very brief look at coupled channels in Dirac space. Later, when discussing the Bethe-Salpeter equation in order to describe mesonic modes within the NJL model, we need to calculate scattering kernels for different Dirac channels, cf. Eq. (3.91). The scattering matrix $\mathcal{T}$ contains a pure Dirac part, $D_M(\Gamma, \Gamma')$, additionally to the structure in flavor and (trivial) color space:

$$\mathcal{T} = -\Gamma_M D_M(\Gamma, \Gamma') \Gamma'_M . \tag{3.32}$$

As summarized in Table 3.3, D_M involves also coupled channels. Since we restrict ourselves to the two-flavor NJL model with scalar and pseudoscalar interaction only, the corresponding submatrix is diagonal in ✓. For instance, introducing also vector and axialvector interactions, this implies a coupled P-A channel. Tensor interactions described by $\sigma^{\mu\nu}$ are not introduced into the NJL model by applying the Fierz transformation to the underlying color-color vector current as seen from Eq. (3.22). However, their presence would lead to coupled-channel effects with the vector and axialvector currents.

	S	P	V	A	T
S	✓	×	✓	×	×
P	×	✓	×	✓	×
V	✓	×	✓	×	✓
A	×	✓	×	✓	✓
T	×	×	✓	✓	✓

Table 3.3.: Coupled-channel analysis in Dirac space: the ✓ denotes the presence of couplings between scalar (S), pseudoscalar (P), vector (V), axialvector (A), or tensor (T) channel, the × denotes that no coupling appears.

3.2. The two-flavor NJL model

The NJL Lagrangian for two flavors contains scalar-interaction and vector-interaction terms only but no anomalous terms from a tensor interaction, $\sigma^{\mu\nu}\sigma_{\mu\nu}$, as it can be seen from Eq. (3.22):

$$\mathcal{L}_{\text{int}}^{\text{2f}} = \frac{G_S}{4}\sum_{i=0}^{3}\left[(\bar{\psi}\tau_i\psi)^2 + (\bar{\psi}i\gamma_5\tau_i\psi)^2\right] - \frac{G_V}{4}\sum_{i=0}^{3}\left[(\bar{\psi}\gamma_\mu\tau_i\psi)^2 + \frac{G_A}{G_V}(\bar{\psi}\gamma_5\gamma_\mu\tau_i\psi)^2\right] \qquad (3.33)$$

In the special case of the color current-current interaction (3.1), the Fierz transformed interaction has only one free coupling parameter: $G_S = 2G_V = 2G_A$. Instead of the full vector interaction, a *reduced vector interaction* is frequently used [Fuk08b, Fuk08a, ZK09]:

$$\tilde{\mathcal{L}}_V = -\frac{G_V}{2}(\bar{\psi}\gamma_\mu\psi)^2 \; . \qquad (3.34)$$

This simplification can be justified a posteriori since it has only small impact on the results. However, as it is demonstrated in [Fuk08b, BHW13] the (full) vector interaction has crucial impacts on the NJL thermodynamics and the resulting phase diagram: both the existence and location of a critical end point (CEP) of the chiral transition are strongly sensitive to the value of G_V. Some more details will be discussed later in Section 3.4.

In the following we use the two-flavors NJL Lagrangian with scalar and pseudoscalar interaction terms only, but with the additional 't Hooft determinant which removes the unwanted $U(1)_A$ symmetry from the Lagrangian. In Euclidean spacetime it reads

$$\mathcal{L}_{\text{NJL}}^{\text{2f}} = \bar{\psi}\left(-i\slashed{\partial}_E + m_0\right)\psi - \frac{G}{2}\left[(\bar{\psi}\psi)^2 + (\bar{\psi}i\gamma_5\boldsymbol{\tau}\psi)^2\right], \qquad (3.35)$$

with the Euclidean γ matrices $\gamma_E^\mu = \gamma_{E,\mu}$ for $\mu \in \{1,2,3,4\}$ in Weyl (chiral) representation:

$$\gamma_0 = \gamma^0 = \begin{pmatrix} 0 & \mathbb{1} \\ \mathbb{1} & 0 \end{pmatrix}, \quad \gamma^i = \begin{pmatrix} 0 & \sigma_i \\ -\sigma_i & 0 \end{pmatrix}, \quad \gamma_4 = i\gamma_0, \quad \gamma_5 = \begin{pmatrix} -\mathbb{1} & 0 \\ 0 & \mathbb{1} \end{pmatrix}. \qquad (3.36)$$

One has $\{\gamma_E^\mu, \gamma_E^\nu\} = -2\delta^{\mu\nu}$ and the *slash notation* defined by

$$\slashed{a}_E = \gamma_E^\mu a_\mu = \gamma_4 a_4 + \boldsymbol{\gamma}\cdot\boldsymbol{a} \; . \qquad (3.37)$$

The Dirac part can be rewritten in Minkowski space,

$$\begin{aligned}\left(-i\slashed{\partial}_E + m_0\right)\psi &= -\left(i\gamma_4\partial_4 + i\boldsymbol{\gamma}\cdot\boldsymbol{\nabla} - m_0\right)\psi = \\ &= -\left(i\gamma^0\partial_0 + i\gamma^i\partial_i - m_0\right)\psi = -\left(i\slashed{\partial} - m_0\right)\psi \; ,\end{aligned} \qquad (3.38)$$

where we have used $\partial_4 = -i\partial_0$ and

$$\partial_\mu = \left(\frac{\partial}{\partial t}, \boldsymbol{\nabla}\right)^T, \quad \partial^\mu = \left(\frac{\partial}{\partial t}, -\boldsymbol{\nabla}\right)^T. \qquad (3.39)$$

The two-flavor NJL Lagrangian reads in Minkowski space:

$$\mathcal{L}_{\text{NJL}}^{\text{2f}} = \bar{\psi}\left(i\slashed{\partial} - m_0\right)\psi + \frac{G}{2}\left[(\bar{\psi}\psi)^2 + (\bar{\psi}i\gamma_5\boldsymbol{\tau}\psi)^2\right]. \qquad (3.40)$$

This is the original version of the NJL Lagrangian as stated in [NJL61a]. In fact, it matches the

3. The Nambu–Jona-Lasinio model

form of the general NJL vertex we concluded in Eq. (3.31) including the 't Hoof determinant:

$$\frac{G}{2}\left[(\bar{\psi}\psi)^2 + (\bar{\psi}i\gamma_5\boldsymbol{\tau}\psi)^2\right] = \\ = \frac{G}{4}\sum_{i=0}^{3}\left[(\bar{\psi}\tau_i\psi)^2 + (\bar{\psi}i\gamma_5\tau_i\psi)^2\right] + \frac{G}{2}\left[\det_{jk}(\bar{\psi}_j(1+\gamma_5)\psi_k) + \det(\bar{\psi}_j(1-\gamma_5)\psi_k)\right], \quad (3.41)$$

where on introduces additionally to the three flavor matrices also $\tau_0 = \mathbb{1}$. In the second line one sums for $\boldsymbol{\tau}^2$ over the three components $i = 1, 2, 3$ only. To be very precise, we write down at this point the two-flavor 't Hooft determinant explicitly for $\psi = (u, d)^{\mathrm{T}}$:

$$\det(\bar{\psi}(1 \pm \gamma_5)\psi) = \bar{u}(1 \pm \gamma_5)u\bar{d}(1 \pm \gamma_5)d - \bar{d}(1 \pm \gamma_5)u\bar{u}(1 \pm \gamma_5)d = \\ = \bar{u}u\bar{d}d \pm (\bar{u}\gamma_5 u\bar{d}d + \bar{u}u\bar{d}\gamma_5 d) - \bar{d}u\bar{u}d \mp (\bar{d}\gamma_5 u\bar{u}d + \bar{d}u\bar{u}\gamma_5 d) + \quad (3.42) \\ + \bar{u}\gamma_5 u\bar{d}\gamma_5 d - \bar{d}\gamma_5 u\bar{u}\gamma_5 d \,.$$

Therefore one finds for the sum of the two determinants

$$\det(\bar{\psi}(1+\gamma_5)\psi) + \det(\bar{\psi}(1-\gamma_5)\psi) = 2(\bar{u}u\bar{d}d - \bar{u}d\bar{d}u) + 2(\bar{u}\gamma_5 u\bar{d}\gamma_5 d - \bar{d}\gamma_5 u\bar{u}\gamma_5 d), \quad (3.43)$$

and a direct comparison of the coefficients of $\mathbb{1}$ and γ_5 confirm the identity (3.41). Apart from the discrete symmetries $\mathcal{C}$, $\mathcal{P}$ and $\mathcal{T}$, the symmetry group of the two-flavor NJL model (3.40) reads in the chiral limit, $m_0 = 0$,

$$\mathrm{SU}(2)_{\mathrm{L}} \times \mathrm{SU}(2)_{\mathrm{R}} \times \mathrm{U}(1)_{\mathrm{V}}, \quad (3.44)$$

where the first two terms denote the chiral symmetry. In the NJL model the axial symmetry, $\mathrm{U}(1)_{\mathrm{A}}$, which is anomalous in QCD, is broken explicitly by the 't Hooft determinant (3.28). However, the primary four-fermion vertex,

$$\sum_{j=0}^{N_f^2-1}\left[(\bar{\psi}T_j\psi)^2 + (\bar{\psi}i\gamma_5 T_j\psi)^2\right], \quad (3.45)$$

is $\mathrm{U}(1)_{\mathrm{A}}$ symmetric, which can be shown using $\mathrm{e}^{2i\gamma_5\alpha} = \cos(2\alpha) + i\gamma_5\sin(2\alpha)$:

$$\left[(\bar{\psi}T_j\psi)^2 + (\bar{\psi}i\gamma_5 T_j\psi)^2\right] \mapsto \left[(\bar{\psi}T_j\mathrm{e}^{2i\gamma_5\alpha}\psi)^2 + (\bar{\psi}i\gamma_5\mathrm{e}^{2i\gamma_5\alpha}T_j\psi)^2\right] = \\ = \left[\bar{\psi}T_j(\cos(2\alpha) + i\gamma_5\sin(2\alpha))\psi\right]^2 + \left[\bar{\psi}T_j(i\gamma_5\cos(2\alpha) - \sin(2\alpha))\psi\right]^2 = \quad (3.46) \\ = \left[(\bar{\psi}T_j\psi)^2 + (\bar{\psi}i\gamma_5 T_j\psi)^2\right].$$

As already stated in the introduction to this chapter, the feature of spontaneous symmetry breaking is the central property of the NJL model. It appears when the coupling parameter G of the two-flavor NJL model (3.40) is sufficiently large. Linearizing the interaction term in the Lagrangian introduces a dynamical quark mass. In Minkowski space one has:

$$\mathcal{L}^{\mathrm{2f,lin}} = \bar{\psi}\left(i\partial\!\!\!/ - m_0\right)\psi + G\big[(\bar{\psi}\psi)\langle\bar{\psi}\psi\rangle + (\bar{\psi}i\gamma_5\boldsymbol{\tau}\psi)\underbrace{\langle\bar{\psi}i\gamma_5\boldsymbol{\tau}\psi\rangle}_{=0}\big]. \quad (3.47)$$

Only the scalar condensate $\langle\bar{\psi}\psi\rangle \neq 0$ can have a non-vanishing vacuum expectation value, the pseudoscalar condensate $\langle\bar{\psi}i\gamma_5\boldsymbol{\tau}\psi\rangle = 0$ must vanish due to the parity symmetry of the (QCD) vacuum. This argument remains true even in an *isotropic* medium with $T, \mu \neq 0$. Only for scenarios where isospin-symmetry breaking takes place, $\mu_{\mathrm{I}} = \mu_{\mathrm{u}} - \mu_{\mathrm{d}} \neq 0$, a non-vanishing pion condensate can form. Anisotropic-medium effects could also induce a non-vanishing pseudoscalar condensate. In this thesis we do not consider such possibilities.

From the linearized Lagrangian (3.47) the effective *constituent-quark mass* can be extracted:

$$m = m_0 - G\langle\bar{\psi}\psi\rangle \,. \tag{3.48}$$

This equation is the so-called *gap equation* (for two flavors), a terminology adapted from the BCS theory of superconductivity. In the chiral limit the constituent-quark mass is just the chiral condensate[23]:

$$\langle\bar{\psi}\psi\rangle = \langle\bar{u}u\rangle + \langle\bar{d}d\rangle + \ldots = -iN_f \, \text{tr} \, G_M^F(0) \,, \tag{3.49}$$

where the trace over the closed quark propagator refers to momentum space and also to Dirac and color space[24]. It is defined as

$$G_M^F(0) = -i \lim_{y \to x^+} \langle\Omega|\mathcal{T}\psi(x)\bar{\psi}(y)|\Omega\rangle \,, \tag{3.50}$$

with the time-ordering symbol $\mathcal{T}$ and the non-perturbative vacuum $|\Omega\rangle$. The momentum integral for (3.49) needs to be regularized which is usually done by a three-momentum cutoff or some Euclidean covariant cutoff. The cutoff $\Lambda \approx 4\pi f_\pi \approx 1$ GeV should be of the order of the chiral scale. We carry out the convergent p_0 integral using residue calculus. Besides the chiral condensate $\langle\bar{\psi}\psi\rangle$ we have also introduced the condensates individually for each quark flavor $\langle\bar{q}q\rangle$, which are defined in the isospin limit as $\langle\bar{\psi}\psi\rangle = N_f \langle\bar{q}q\rangle$. Therefore we have

$$\langle\bar{q}q\rangle = -i \, \text{tr} \int \frac{d^4p}{(2\pi)^4} \frac{1}{\not{p} - m + i\epsilon} = -i \, N_c \int \frac{dp_0}{2\pi} \int \frac{d^3p}{(2\pi)^3} \frac{4m}{p_0^2 - (\mathbf{p}^2 + m^2 - i\epsilon)} =$$
$$\stackrel{\text{Res}}{=} -N_c \int \frac{d^3p}{(2\pi)^3} \frac{2m}{\sqrt{\mathbf{p}^2 + m^2}} = -\frac{m N_c}{2\pi^2} \left(\Lambda\sqrt{m^2 + \Lambda^2} - m^2 \, \text{Arsinh}\left(\frac{\Lambda}{m}\right) \right) \,. \tag{3.51}$$

The chiral condensate is always negative and diverges quadratically when the cutoff Λ becomes large. Therefore, one can conclude from (3.48) that $m \geq m_0$ and arrives at the self-consistent gap equation which needs to be solved for the dynamical constituent-quark mass m:

$$\frac{m - m_0}{m} = \frac{G N_c}{\pi^2} \left(\Lambda\sqrt{m^2 + \Lambda^2} - m^2 \, \text{Arsinh}\left(\frac{\Lambda}{m}\right) \right) \,. \tag{3.52}$$

For $m_0 \neq 0$ there is always a solution with some $m > m_0$. In the chiral limit, $m_0 = 0$, the right hand side is bounded from above, therefore

$$1 \leq \frac{G N_c \Lambda^2}{\pi^2} \,, \tag{3.53}$$

which defines a critical coupling strength G_{cr} and the corresponding dimensionless coupling g_{cr} and the critical NJL fine structure constant α_{cr}:

$$G_{\text{cr}} = \frac{\pi^2}{N_c \Lambda^2} \quad \Rightarrow \quad g_{\text{cr}}^2 = G_{\text{cr}} \Lambda^2 \quad \text{and} \quad \alpha_{\text{cr}} = \frac{g_{\text{cr}}^2}{4\pi} \,. \tag{3.54}$$

---

[23] At this stage we just want to note that there is some ambiguity in the usage of *chiral condensate*. Additionally to the one defined by $\langle\bar{\psi}\psi\rangle$, one can introduce a subtracted version denoted by $\langle:\bar{\psi}\psi:\rangle$ as we will do later in Eq. (3.84). By construction this subtracted condensate vanishes in the limit $m \to m_0$, independent of T and μ. We refer to the more detailed discussion in Section 3.4.

[24] The factor N_f arises since the condensates $\langle\bar{u}u\rangle$, $\langle\bar{d}d\rangle$, ... coincide in the isospin limit. However, for $N_f > 2$ the functional structure of the gap equation (3.48) changes due to the 't Hooft interaction, therefore, this relation between m and $\langle\bar{\psi}\psi\rangle$ is only true for $N_c = 2$ and can not be generalized.

3. The Nambu–Jona-Lasinio model

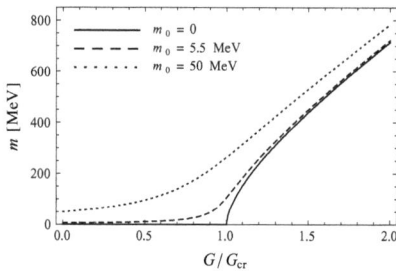
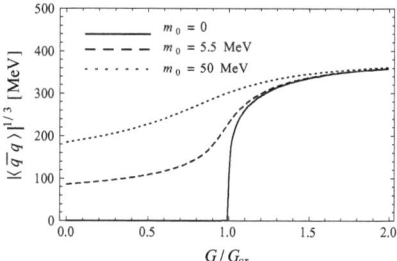

(a) Constituent-quark mass as solution of the vacuum gap equation for different values of the current-quark mass m_0

(b) One-flavor chiral condensate $\langle \bar{q}q \rangle$ for different values of the current-quark mass m_0

Figure 3.2.: Spontaneous chiral symmetry breaking: constituent-quark masses and chiral condensate as function of the NJL four-vertex coupling G with the NJL cutoff $\Lambda = 651$ MeV

Whereas the coupling parameter in the NJL model has the dimension length)^2, or mass dimension $[G] = -2$, the rescaled coupling $g = G\Lambda^2$ is just a number. In the large-N_c limit the critical coupling $G_{\text{cr}} \sim 1/N_c$ becomes small, hence spontaneous chiral symmetry is present also for very small numerical values of the four-fermion coupling G. It is interesting to observe the same behavior of $G_{\text{cr}} \sim 1/\Lambda^2$ for large values of the cutoff, meaning that the chosen regularization scheme affects the feature of spontaneous chiral symmetry breaking within the NJL model.

Solutions of the vacuum gap equation (3.52) are shown in Fig. 3.2(a). It can be seen that spontaneous chiral symmetry breaking can happen in the chiral limit only for $G > G_{\text{cr}}$, whereas for some finite $m_0 > 0$ a solution $m > m_0$ can be found also for arbitrarily small values of G. With the constituent-quark masses one calculates the (one-flavor) chiral condensate $\langle \bar{q}q \rangle$ (3.51) as shown in Fig. 3.2(b) for different values of the current-quark mass m_0.

3.3. Large-N_c analysis for the NJL model

3.3.1. Topology of connected QCD vertices

All interaction terms in the NJL model are vertices with N external fermion pairs, denoted by the local interaction kernels K_{2N}. These vertices originate from QCD after integrating out all gluonic degrees of freedom. Given such a vertex, the building blocks for its internal non-perturbative structure are four-gluon vertices (V_4), three-gluon vertices (V_3), and quark-gluon vertices (V_g). Since the effective $2N$-vertex contains connected diagrams only we have[25] $V_g \geq N$. By "connected" we mean that it is possible to get from an arbitrary external leg to any other external leg just following internal lines and vertices. Assuming that this would not be possible for some internal QCD process, then the vertex can be factorized into (at least) two vertices with fewer external legs:

$$K_{2N} = K_{2N_1} \otimes K_{2N_2} , \qquad (3.55)$$

with $N = N_1 + N_2$. This means that the assumed disconnected QCD process does not contribute to the genuine $2N$-vertex. We illustrate this general statement by two examples: For the four-

---

[25] To be precise, for this statement one has to exclude the free propagator, $N = 1$, without any radiation or interaction.

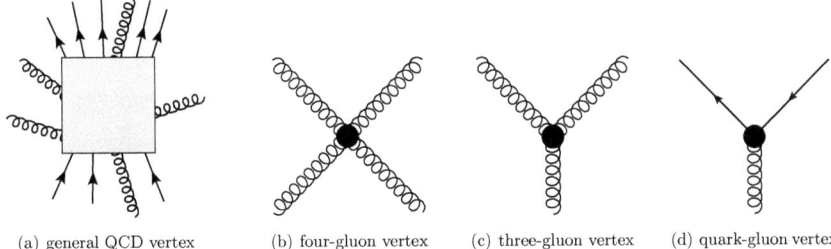

(a) general QCD vertex   (b) four-gluon vertex   (c) three-gluon vertex   (d) quark-gluon vertex

Figure 3.3.: Sketch of a general QCD vertex (a), here with $2N = 8$ external quarks and $B = 5$ external gluons. Its internal structure in terms of QCD vertices (b)-(d), counted by V_4, V_3, B_g, and the number of loops, L, is topologically restricted by (3.56).

vertex a disconnected process would imply $K_4 = K_2 \otimes K_2$ meaning that this contribution to the four-fermion interaction can be described by two independent full quark propagators. It is obvious that this type of non-interacting (skeleton) diagrams does not contribute to the genuine four-fermion vertex. We can carry out a similar argument also for the six-fermion vertex in the three-flavor case: Assuming a non-connected process inside the 't Hooft vertex leads to either $K_6 = K_2 \otimes K_2 \otimes K_2$ or $K_6 = K_4 \otimes K_2$. The first case describes three independent but dressed quark propagators and the second one describes a four-fermion vertex which is "observed" by a full quark propagator. In both cases the disconnected process does not contribute to a genuine six-fermion vertex.

The connected topology of the $2N$-vertices leads to the following identity:

$$V_4 + \frac{1}{2}(V_3 + V_g) = L + N - 1 + \frac{B}{2}, \quad (3.56)$$

where L denotes the number of (momentum) loops in the internal process and B the number of external gluons (gauge bosons) which we state for completeness, cf. Fig. 3.3(a). A formal proof of this equation can be done by induction over N and B. The left-hand side the topological constraint in Eq. (3.56) contains QCD dynamics and introduces a N_c counting to the NJL model. Since the four-gluon vertex is proportional to g_{QCD}^2 but the three-gluon and quark-gluon vertex scale linearly with the QCD coupling, the considered $2N$ vertex scales as

$$K_{2N} \sim \left(g_{\text{QCD}}^2\right)^{V_4} \cdot g_{\text{QCD}}^{V_3+V_g} \cdot N_c^l =$$
$$= g_{\text{QCD}}^{2[V_4+\frac{1}{2}(V_3+V_g)]} \cdot N_c^l \sim \quad (3.57)$$
$$\sim N_c^{l-(L+N-1)},$$

where we have set $B = 0$ in the NJL model and have used the N_c-scaling of the strong coupling constant, $\alpha_s \sim g_{\text{QCD}}^2 \sim 1/N_c$ as it can be seen from the QCD beta function. We denote the number of closed color loops (color traces) by $l \leq L$, each giving rise to an additional factor of N_c in the scaling. Since QCD is a non-Abelian gauge theory where gauge bosons can couple to themselves, there is for any combination of external legs a tree-level diagram without any loop, i.e. $l = L = 0$. We emphasize that this statement is wrong in an Abelian gauge theory: in QED, the leading-order contribution to photon-photon scattering, $\gamma\gamma \to \gamma\gamma$, is given by an electron-loop diagram of order α_{QED}^2 and due to the absence of a multi-photon vertex no tree-level diagram contributes.

3. The Nambu–Jona-Lasinio model

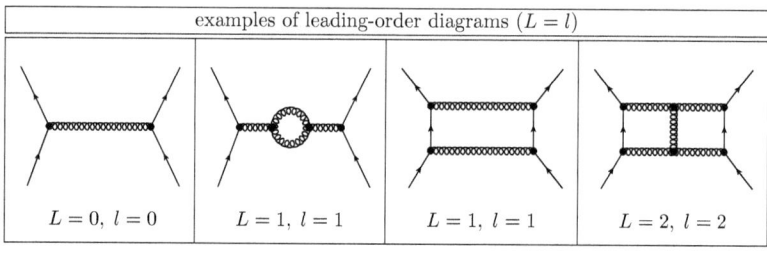

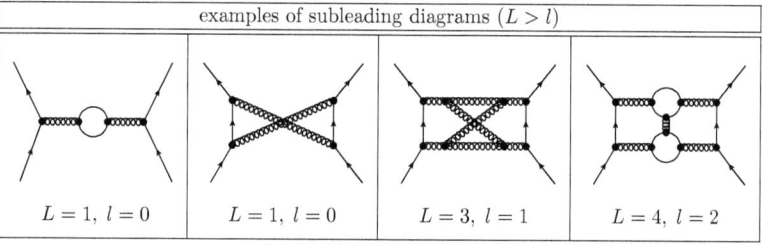

Table 3.4.: Examples of QCD processes contributing to the effective NJL four-fermion vertex $K_4 = G \sim 1/N_c$ at leading order (upper panel with $L = l$) and at subleading orders (lower panel with $L > l$).

We conclude that the leading-order and simplest (i.e. tree level) contribution to a $2N$-vertex in the NJL model scales as:

$$K_{2N} \sim \frac{1}{N_c^{N-1}} \quad \Rightarrow \quad G = K_4 \sim \frac{1}{N_c} \,. \tag{3.58}$$

We emphasize that this is a non-perturbative result since in addition to tree-level diagrams all QCD processes at all orders in α_s with $L = l$, i.e. all planar diagrams without closed internal Fermi lines [tH74a, tH74b, Wit79], contribute to the effective NJL vertices. In Table 3.4 we show a few examples of leading-order and subleading-order diagrams contributing to the four-fermion vertex G.

The large-N_c scaling of K_{2N} is of crucial importance for our treatment of the NJL model. We will use this scaling of the coupling as a bookkeeping method to organize the non-perturbative NJL model in a systematic way. However, despite applying a large-N_c counting method we do not use simple perturbation theory in a small coupling $G \sim 1/N_c$. As we will develop in the following, already at next-to-leading order to the vacuum partition function in $1/N_c$, a resummation of all orders in the coupling G must be performed.

We conclude this discussion by commenting briefly on a conceptual difference between the NJL model and QCD as underlying theory. The large-N_c scaling we have just presented applies to any K_{2N} vertex in the NJL model including the 't Hooft determinant K_6 in the three-flavor case. The determinant (3.28) is added to the NJL Lagrangian in order to remove the unwanted $U(1)_A$-symmetry: in the NJL model this symmetry is broken explicitly, whereas in QCD it is broken anomalously. The instanton vacuum in QCD is not described by the non-perturbative but still diagrammatic treatment of the internal structure of NJL vertices. In the following we review briefly how instantons give rise to anomalous symmetry breaking and why such effects cannot be treated using perturbative approaches.

Linearly divergent triangle diagrams which express a non-trivial transformation of the path-integral measure under the U(1)$_A$-symmetry show that quantization of the QCD Lagrangian breaks the classical U(1)$_A$-symmetry [Adl69, Fuj79]. This phenomenon can be explained by instantons in the QCD vacuum characterized by the topological quantum number $q \in \mathbb{N}$ (winding number)[CL06]:

$$q = \frac{g_{\text{QCD}}^2}{32\pi^2} \int \mathrm{d}^4 x \, G_{\mu\nu}^a \widetilde{G}^{\mu\nu,a} \,, \tag{3.59}$$

with the dual gluonic field-strength tensor

$$\widetilde{G}^{\mu\nu,a} = \frac{1}{2}\epsilon^{\mu\nu\alpha\beta} G_{\alpha\beta}^a \,. \tag{3.60}$$

Having a pure Yang-Mills theory in mind, any instanton solution is either self-dual or anti-self-dual: $\widetilde{G} = \pm G$. One therefore derives:

$$\exp\left(\int \mathrm{d}^4 x \, \mathcal{L}\right) = \exp\left(\pm \frac{1}{4} \int \mathrm{d}^4 x \, G_{\mu\nu} \widetilde{G}^{\mu\nu}\right) = \exp\left(\pm \frac{8\pi^2 q}{g_{\text{QCD}}^2}\right). \tag{3.61}$$

From this expression it can be seen that instantons are purely non-perturbative in terms of an expansion in a small coupling parameter g_{QCD}. However, they introduce a mechanism which violates the axial-charge conservation [tH76], which can be understood when comparing with the divergence of the axialvector current $j_5^\mu = \bar{\psi}\gamma_\mu\gamma_5\psi$, which is non-zero even in the chiral limit[CL06]:

$$\partial_\mu j_5^\mu = -\frac{2 N_{\text{f}} g_{\text{QCD}}^2}{32\pi^2} G_{\mu\nu}^a \widetilde{G}^{\mu\nu,a} \,. \tag{3.62}$$

The corresponding change in the axialvector charge is directly related to the winding number q, which is therefore also denoted as topological charge of the instanton[26]:

$$\Delta Q_5 = \int \mathrm{d}^4 x \, \partial_\mu j_5^\mu = -2 N_{\text{f}} q \,. \tag{3.63}$$

From this we see that an instanton in the QCD vacuum can change the axialvector charge from $+1$ to -1 for each flavor and each topological-charge unit. This shows the violation of the axial-charge conservation and the anomalous breaking of the classical U(1)$_A$-symmetry of QCD.

3.3.2. Generating functional of the NJL model

Next we follow [MBW10] and construct a thermal two-particle irreducible (2PI) generating functional Φ from which we can derive $2N$-point functions relevant for both the quark and meson sector. We switch from Minkowski to Matsubara space (cf. Appendix A.2) and denote the full thermal quark propagator by G_β^{F}. The bookkeeping is done in terms of the N_{c}-scaling since the potential Φ is fully non-perturbative:

$$\Phi = \sum_{k=0}^{\infty} \Phi^{(k)} \,, \quad \text{where} \quad \Phi^{(k)} \sim N_{\text{c}}^{1-k} \,. \tag{3.64}$$

From this, using functional calculus, the quark self-energy and the interaction kernel, i.e. the

---

[26]The trivial solution of the Yang-Mills equation feature $q = 0$ and in most applications the single-charged instanton solution with $q = 1$ is relevant, in particular for the discussion in [tH76].

3. The Nambu–Jona-Lasinio model

four vertex, can be calculated when differentiating with respect to full quark propagators[27]:

$$\Sigma^{(k)} = \frac{\delta \Phi^{(k)}}{\delta G_\beta^F}, \qquad K^{(k)} = \frac{\delta^2 \Phi^{(k)}}{(\delta G_\beta^F)(\delta G_\beta^F)}. \tag{3.65}$$

Only closed (vacuum) diagrams are summed in the potential Φ, therefore the number of fermion loops, l, and the number of NJL vertices, n, are restricted by $l \leq L = n+1$, where L denotes as before the number of loops w.r.t. to the momentum integration. This shows that the leading-order term in the potential is $\Phi^{(0)}$ which scales with $N_c^{-n} \cdot N_c^l \leq N_c^1$. At this order there is only one (Hartree) diagram contributing to the potential:

$$\Phi^{(0)} = \quad \bigcirc\!\!\bullet\!\!\bullet\!\!\bigcirc \tag{3.66}$$

In order to distinguish between Hartree and Fock contributions to the four-quark coupling G we have introduced the tiny wavy line indicating which quark loop refers to a color trace. Due to the restriction $l = n+1$ any new pair of vertex and fermion loop leads already to a two-particle reducible diagrams which imply disconnected contributions to $K^{(0)}$. Therefore, the leading-order potential is given by a single vacuum diagram. It scales as $\Phi^{(0)} \sim N_c$.

At next-to-leading order with $l = n$ there are already infinitely many diagrams:

$$\Phi_M^{(1)} = \quad \text{(diagrams)} \quad + \quad \text{(diagrams)} \quad + \quad \text{(diagrams)} \quad + \quad \text{(diagrams)} \quad + \ldots \tag{3.67}$$

Including the global symmetry factor of $1/2$ the diagrammatic expression at leading order reads

$$\Phi^{(0)} = \frac{G}{2} \left(T \sum_{n \in \mathbb{Z}} \int \frac{d^3 q}{(2\pi)^3} \operatorname{Tr} G_\beta^F(\boldsymbol{q}, \nu_n) \right)^2. \tag{3.68}$$

The generating functional $\Phi^{(0)}$ does not carry a channel index M, since only $\Gamma = \mathbb{1}$ leads to a non-vanishing diagram. In the pseudoscalar channel the traceless γ_5 matrix leads to a vanishing diagram.

For $\Phi^{(1)}$ one needs to specify additionally the channel, where $\Gamma_M \in \{\mathbb{1}, i\gamma_5\}$ describes the scalar and pseudoscalar case, respectively:

$$\begin{aligned}\Phi_M^{(1)} &= \frac{1}{2} T \sum_{n \in \mathbb{Z}} \int \frac{d^3 p}{(2\pi)^3} \sum_{k=1}^\infty \frac{1}{k} G^k \Pi_M^k(\boldsymbol{p}, \omega_n) = \\ &= -\frac{1}{2} T \sum_{n \in \mathbb{Z}} \int \frac{d^3 p}{(2\pi)^3} \ln\left[1 - G \Pi_M(\boldsymbol{p}, \omega_n)\right],\end{aligned} \tag{3.69}$$

---

[27]Note the dimension of the functional derivative:

$$\left[\frac{\delta^n f(x)}{\delta f_1(x_1) \ldots \delta f_k(x_k)}\right] = [f] - \sum_{i=1}^k \left([f_i] + \dim(x_i)\,[x_i]\right),$$

where $\dim(x) = n$ if $x \in \mathbb{K}^n$ is a vector in a n-dimensional vector space over $K \in \{\mathbb{R}, \mathbb{C}\}$. From this the correct mass dimensions are ensured: $[\Sigma^{(k)}] = 1$ and $[K^{(k)}] = -2$ with $[\Phi] = 4$.

where Π_M denotes the quark-antiquark loop (polarization tensor)

$$\Pi_M(\boldsymbol{p},\omega_n) = -T \sum_{m\in\mathbb{Z}} \int \frac{\mathrm{d}^3 q}{(2\pi)^3} \mathrm{Tr}\left[\Gamma_M G_\beta^F(\boldsymbol{q},\nu_m)\Gamma_M G_\beta^F(\boldsymbol{q}-\boldsymbol{p},\nu_m-\omega_n)\right]. \quad (3.70)$$

From the generating functional we can derive all self-energy insertions and kernels which are present up to the considered order in $1/N_c$. Diagrammatically the functional derivative $\delta/\delta G_\beta^F$ means to cut one closed quark line in the diagram:

$$\Sigma^{(0)} = \frac{\delta \Phi^{(0)}}{\delta G_\beta^F} = \cdots = GT \sum_{n\in\mathbb{Z}} \int \frac{\mathrm{d}^3 q}{(2\pi)^3} \mathrm{Tr}\, G_\beta^F(\boldsymbol{q},\nu_n),$$

$$\Sigma_M^{(1)}(\boldsymbol{p},\nu_n) = \frac{\delta \Phi^{(1)}}{\delta G_\beta^F} = \cdots = -T \sum_{m\in\mathbb{Z}} \int \frac{\mathrm{d}^3 q}{(2\pi)^3} D_M(\boldsymbol{q},\omega_m) \mathrm{Tr}\left[\Gamma_M^2 G_\beta^F(\boldsymbol{q}-\boldsymbol{p},\nu_m-\omega_n)\right],$$
(3.71)

where we have introduced in $\Sigma_M^{(1)}$ a bosonic propagator arising from the functional derivative of $\ln[1-G\Pi_M(\boldsymbol{q},\omega_n)]$ in the integrand of $\Phi^{(1)}$:

$$D_M(\boldsymbol{q},\omega_m) = \frac{G}{1-G\Pi_M(\boldsymbol{q},\omega_m)}. \quad (3.72)$$

In the next section when discussing the Bethe-Salpeter equation we will notice that this bosonic propagator is nothing but the renormalized full meson propagator, cf. Eq. (3.98). The propagators for the pion and the sigma boson are denoted as D_π and D_σ, arising from the pseudoscalar and scalar channel, respectively.

From the partition function also the four-fermion interaction kernel K can be derived by applying once more the functional derivative with respect to G_β^F:

$$K^{(0)} = \frac{\delta \Sigma^{(0)}}{\delta G_\beta^F} = \cdots = G,$$

$$K_M^{(1)}(\boldsymbol{p},\nu_n) = \frac{\delta \Sigma_M^{(1)}}{\delta G_\beta^F} = \cdots + \mathcal{O}(N_c^{-2}) = -\Gamma_M D_M(\boldsymbol{p},\omega_n)\Gamma_M + \mathcal{O}(N_c^{-2}).$$
(3.73)

Note that there are higher-order corrections to the meson kernel $K_M^{(1)}$ which arise from differentiating in the integrand once again D_M with respect to G_β^F. Diagrammatically this means that the second functional derivative cuts a quark line of a loop that is different from the first one. Therefore, two different topologies arise when applying $\delta^2/(\delta G_\beta^F)^2$ to $\Phi_M^{(1)}$:

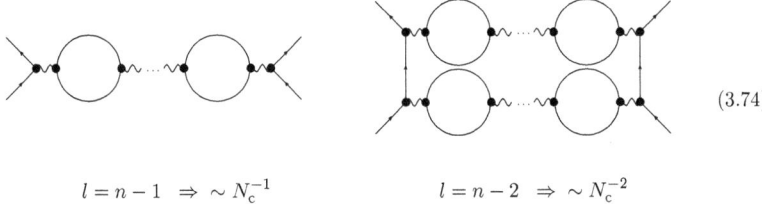

$$l = n-1 \;\Rightarrow\; \sim N_c^{-1} \qquad\qquad l = n-2 \;\Rightarrow\; \sim N_c^{-2}$$
(3.74)

Having performed all functional derivatives we know all building blocks at next-to-leading order in a $1/N_c$-expansion relevant to construct the gap equation (quark 2-point function) and Bethe-Salpeter equation (quark 4-point function). This means that applying a large-N_c analysis to the

3. The Nambu–Jona-Lasinio model

NJL generating functional one rediscovers standard approaches used in many-particle physics as described in standard textbooks, e.g. [Mat76, FW03], and in particular in the NJL model itself [KLVW90a, KLVW90b]. At leading order in large-N_c we find the *gap equation*,

$$\mathcal{O}(1): \quad \text{———■———} \quad = \quad \text{———} \quad + \quad \text{—◯—■—} \tag{3.75}$$

and at next-to-leading order the *Bethe-Salpeter equation*,

$$\mathcal{O}(N_c^{-1}): \quad \text{>=<} \quad = \quad \text{>•<} \quad + \quad \text{>•◯•=<} \tag{3.76}$$

These two equations represent the first two Dyson-Schwinger equations which can be derived from the generating functional Φ. Due to their non-perturbative nature, both the gap equation and Bethe-Salpeter equation (BSE) are self-consistent equations including full constituent-quark and meson propagators. The resummation of quark-antiquark loops leads to mesonic modes and the solutions of this self-consistent equation can be interpreted as renormalized meson propagators. We note that the BSE is considered in the so-called random-phase approximation (RPA) which means that only "resonant" quark-antiquark loops are coherently resummed to all orders. In principle there are also the t-channel and u-channel contributions to the BSE at order $1/N_c$, but the mesonic fluctuation $\Sigma_M^{(1)}$ includes only mesonic modes from the quark-antiquark s-channel. For self-consistency reasons we therefore truncate the BSE also to this channel.

The mesonic fluctuations described by $\Sigma_M^{(1)}$ in Eq. (3.71) contribute at next-to-leading order to the gap equation (3.75):

$$\text{———■———} \quad = \quad \text{———} \quad + \quad \text{—◯—■—} \quad + \quad \text{—⌒⌒—} \tag{3.77}$$

The fluctuation term leads to a coupling between the quark and the meson sector and spoils the analytical approach to simultaneously exact solutions of both equations. In principle, this term also introduces a momentum-dependent quark mass as it is derived from QCD Dyson-Schwinger calculations. We discuss the two equations separately and neglect the coupling between them. In particular no momentum dependence of the constituent-quark mass is taken into account. One can justify this by arguing that the mesonic fluctuations act only as Fock exchange corrections of order $\mathcal{O}(N_c^{-1})$ to the actual Hartree gap equation.

The NJL model incorporates the mesonic modes by introducing additional effective Yukawa terms in the Lagrangian[Kle92, Bub05]. For example, the coupling of the pion to the quark is described by

$$\mathcal{L}_{\pi qq} = -g_{\pi qq} \bar{\psi} i\gamma_5 \boldsymbol{\tau} \psi \cdot \boldsymbol{\pi} + \frac{f_{\pi qq}}{m_\pi} \bar{\psi} \gamma_\mu \gamma_5 \boldsymbol{\tau} \psi \cdot \partial^\mu \boldsymbol{\pi} \ , \tag{3.78}$$

where we have introduced two couplings arising as residua of the pion propagator D_π, cf. the detailed discussion in Section 3.5.2. In this thesis we consider the case of vanishing axialvector coupling $f_{\pi qq} = 0$ only. The corresponding Yukawa coupling of the sigma boson is:

$$\mathcal{L}_{\sigma qq} = -g_{\sigma qq} \bar{\psi} \sigma \psi \ , \tag{3.79}$$

where the coupling $g_{\sigma qq}$ is also derived as wave-function renormalization constant from the corresponding mesonic propagator. In the chiral limit, the two quark-meson couplings coincide, $g_{\sigma qq} = g_{\pi qq}$. This is approximatively fulfilled also for high temperatures $T \gtrsim 200$ MeV.

3.4. Gap equation and thermal constituent-quark masses

In the NJL model the (thermal) constituent-quark masses are generated dynamically by spontaneous chiral symmetry breaking. This is described by the gap equation (3.75) which we truncate to its Hartree level for the time being. The mesonic Fock terms present in Eq. (3.77) are not considered in this approximation. In the two-flavor case the gap equation reads $m = m_0 - G\langle\bar{\psi}\psi\rangle$ which results from linearizing the NJL Lagrangian in Eq. (3.48). In order to describe not only the vacuum constituent-quark mass but also thermal quark masses we use the Matsubara formalism and switch from Minkowski to Matsubara space (cf. Appendix A.2):

$$i \int \frac{\mathrm{d}^4 p}{(2\pi)^4} \, I(p) \mapsto T \sum_{n \in \mathbb{Z}} \int \frac{\mathrm{d}^3 p}{(2\pi)^3} \, I(\boldsymbol{p}, p_n) \,. \tag{3.80}$$

The thermal chiral condensate is therefore given by

$$\begin{aligned}
\langle \bar{q} q \rangle &= -\mathrm{tr}\, T \sum_{n \in \mathbb{Z}} \int \frac{\mathrm{d}^3 q}{(2\pi)^3} \frac{\nu_n \gamma_4 - \boldsymbol{p} \cdot \boldsymbol{\gamma} + m}{\nu_n^2 + \boldsymbol{p}^2 + m^2} = \\
&= -4 N_c T \sum_{n \in \mathbb{Z}} \int \frac{\mathrm{d}^3 p}{(2\pi)^3} \frac{m}{\nu_n^2 + E_p^2} = \\
&= -\frac{4 N_c \cdot 4\pi}{8\pi^3} \int_0^\Lambda \mathrm{d}p \, \frac{p^2 m}{2 E_p} \left(1 - n_{\mathrm{F}}^+(E_p) - n_{\mathrm{F}}^-(E_p) \right) \,,
\end{aligned} \tag{3.81}$$

where we used the fermionic Matsubara frequencies $\nu_n = (2n+1)\pi T - \mathrm{i}\mu$. We have introduced $E_p = \sqrt{\boldsymbol{p}^2 + m^2}$ and $n_{\mathrm{F}}^\pm$ denoting the Fermi distribution function describing thermal quarks and antiquarks, respectively:

$$n_{\mathrm{F}}^\pm(E_p) = \frac{1}{e^{\beta(E_p \mp \mu)} + 1} \,. \tag{3.82}$$

The thermal gap equation reads

$$m = m_0 - 2G\langle \bar{q}q \rangle = m_0 + \frac{2 G N_c m}{\pi^2} \int_0^\Lambda \mathrm{d}p \, \frac{p^2}{E_p} \left(1 - n_{\mathrm{F}}^+(E_p) - n_{\mathrm{F}}^-(E_p) \right) \,. \tag{3.83}$$

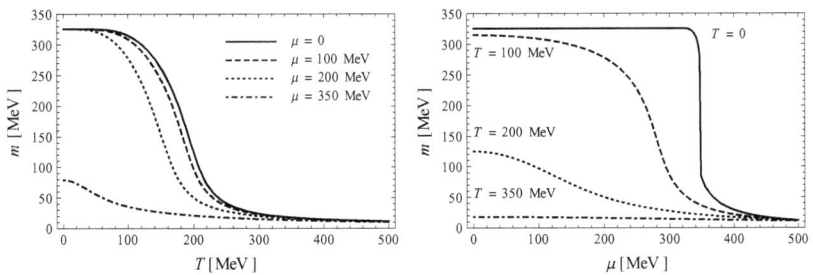

(a) T-dependence of quark mass for different values of the quark chemical potential μ

(b) μ-dependence of quark mass for different values of the temperature T

Figure 3.4.: Thermal constituent-quark masses $m(T, \mu)$ from self-consistent solutions of the leading-order gap equation (3.75) (Hartree level)

3. The Nambu–Jona-Lasinio model

Comparing this result to that in Minkowski space in Eq. (3.51) one rediscovers the vacuum result when sending first $\mu \to 0$ and then $T \to 0$. This is because $n_F^+(E) \to 0$ for $T \to 0$ only if $E > \mu$. The cutoff Λ is introduced to regularize the divergent part of the chiral condensate $\langle \bar{q}q \rangle$ in Eq. (3.81). Note that the thermal distribution functions could be integrated also for $p > \Lambda$, but there is no physical contributions to the condensate for large momenta. However, the condensate itself is not a physical observable. It is usually grouped together with the NJL coupling G which is sometimes denoted as $\sigma = G\langle \bar{q}q \rangle$. For large momenta the NJL coupling vanishes and the quarks are non-interacting: $G(p > \Lambda) = 0$.

The results for the constituent-quark mass as function of temperature and quark chemical potential are shown in Fig. 3.4. The NJL parameters (m_0, G, Λ) are chosen in such a way that in the vacuum the constituent-quark mass m, the pion mass m_π and the pion decay constant f_π reproduce physical values. The mesonic sector is only discussed in the next section. We summarize our NJL parameter set in Table 3.5.

Using these parameters we find a monotonically decreasing constituent-quark mass as function of T and μ. Defining the coordinates (T, μ) where the quark mass drops to half its vacuum value, $m(T, \mu) = \frac{1}{2} m(0,0)$, we construct the phase diagram of the two-flavor NJL model shown in Fig. 3.5 defining the chiral crossover/transition line $T_\chi(\mu)$. For large quark chemical potentials there is a first-order transition line which ends in a critical endpoint localized at $(T, \mu)_{\text{CEP}} = (43 \text{ MeV}, 330 \text{ MeV})$. For $\mu < \mu_{\text{CEP}}$ there is no actual phase transition anymore, but only a smooth crossover transition. For high temperatures the quark condensate melts down to small values, hence chiral symmetry is said to be in the Wigner-Weyl realization where $Q_a^A |0\rangle = 0$. In contrast, close to the vacuum at low T and μ chiral symmetry is spontaneously broken and said to be in the Nambu-Goldstone realization where $Q_a^A |0\ rangle \neq 0$. Here Q_a^A denotes the axial charge operator arising from $A_\mu^a = \bar{\psi}\gamma_\mu\gamma_5 T^a \psi$, where T^a are the generators of the SU(N_f) flavor group. We note that not only the chiral condensate can break chiral symmetry but also diquark condensates, $\langle \psi\psi \rangle$, which can be realized at high densities, i.e. high quark chemical potentials. Such a scenario is not considered in this work. As we will discuss in detail in Section 3.7 the respective degrees of freedom feature large masses and do not give a significant contribution to the shear viscosity.

We conclude this discussion with a more detailed look at the chiral condensate which can also be defined in a subtracted version (in Matsubara space):

$$\langle :\bar{\psi}\psi: \rangle = -N_f \left(\text{tr} \, G_\beta^F(0) - \text{tr} \, G_\beta^F(0)\big|_{m \to m_0} \right), \qquad (3.84)$$

where $G_\beta^F(0)$ denotes the closed fermion propagator in position space, cf. Eq. (3.50). As in Minkowski space the trace covers momentum, Dirac, and color space. In the chiral limit both the subtracted and the non-subtracted chiral condensate coincide. One finds in general

$$\lim_{m \to m_0} |\langle \bar{\psi}\psi \rangle| > \lim_{m \to m_0} |\langle :\bar{\psi}\psi: \rangle| = 0, \qquad (3.85)$$

Input			Output [MeV]				
m_0	G	Λ	m	m_π	f_π	$\langle \bar{\psi}\psi \rangle^{1/3}$	$\langle \bar{q}q \rangle^{1/3}$
5.50 MeV	10.08 GeV^{-2}	651 MeV	325	140	92.4	-316	-251

Table 3.5.: Summary of our model parameters and vacuum results in the NJL quark and meson sector discussed in the next section. The input parameter set is taken from [RÖ6]

3.4. Gap equation and thermal constituent-quark masses

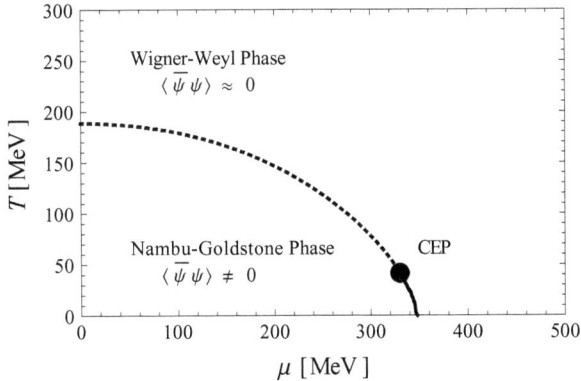

Figure 3.5.: Phase diagram of the two-flavor NJL model. The dotted line denotes the crossover temperature as function of the quark chemical potential separating the Nambu-Goldstone phase where chiral symmetry is spontaneously and the Wigner-Weyl phase where chiral symmetry is restored. The small solid line of first-order phase transitions ends at a critical end point (CEP) which is located at unphysically low temperatures which is known to be a shortcoming of the NJL model we use in this work.

compare also Fig. 3.2(b). For a finite but small current-quark mass, $m_0 \neq 0$, the subtracted chiral condensate can still be interpreted as an order parameter for spontaneous chiral symmetry breaking despite the fact that in this case chiral symmetry is only an approximate symmetry of the NJL Lagrangian. We have denoted the subtracted chiral condensate by two colons because this condensate refers to the normal-ordered product of quark and antiquark fields. Denoting the *perturbative* (QCD) vacuum by $|0\rangle$ we get from Wick's Theorem [PS95]:

$$\mathcal{T}\psi\bar{\psi} = :\psi\bar{\psi}: + \langle 0|\mathcal{T}\psi\bar{\psi}|0\rangle , \qquad (3.86)$$

where $\mathcal{T}$ stands for the time-ordering symbol and $::$ for the normal-ordering symbol. Sandwiching this identify between the perturbative vacuum we find $\langle 0|:\psi\bar{\psi}:|0\rangle = 0$. Only in the *non-perturbative* vacuum $|\Omega\rangle$ a non-vanishing condensate can form:

$$\begin{aligned}\langle:\bar{\psi}\psi:\rangle &= -\mathrm{Tr}\,\langle\Omega|:\psi\bar{\psi}:|\Omega\rangle = \\ &= -N_\mathrm{f}\,\mathrm{tr}\left(\langle\Omega|\mathcal{T}q\bar{q}|\Omega\rangle - \langle 0|\mathcal{T}q\bar{q}|0\rangle\right),\end{aligned} \qquad (3.87)$$

which coincides with the definition of the subtracted condensate in Eq. (3.84). In Feynman diagrams only time-ordered correlators are encoded, hence we identify the (non-subtracted) chiral condensate that enters into the gap equation (3.83):

$$\langle\bar{\psi}\psi\rangle = -N_\mathrm{f}\,\mathrm{tr}\,\langle\Omega|\mathcal{T}q\bar{q}|\Omega\rangle = -N_\mathrm{f}\,\mathrm{tr}\,G^\mathrm{F}_\beta(0) . \qquad (3.88)$$

As a last comment in this section we consider the decomposition of the constituent-quark mass,

$$m = m_0 + \bar{m}_0 + \bar{m} \to \begin{cases} \bar{m} & \text{for } m_0 \to 0 \\ m_0 & \text{for } T \to \infty \end{cases}, \qquad (3.89)$$

3. The Nambu–Jona-Lasinio model

in free, perturbative and non-perturbative contributions, respectively. Then we have in the chiral limit:

$$m = \bar{m} = 0 \quad \Leftrightarrow \quad \langle \bar{\psi}\psi \rangle = \langle :\bar{\psi}\psi: \rangle = 0 \,. \tag{3.90}$$

Due to conservation of chirality (helicity) at all (QCD) vertices a non-zero quark mass cannot be generated perturbatively in the chiral limit.

3.5. Bethe-Salpeter equation and thermal meson masses

3.5.1. Meson masses and mesonic spectral functions

The mesonic (soft) modes are described by the Bethe-Salpeter equation (3.76): a resummation of ring diagrams made of quark-antiquark loops introduce the next-to-leading order Dyson-Schwinger equation in a large-N_c expansion. The general structure of the quark-antiquark loop (polarization tensor) for both Dirac channels, $\Gamma \in \{\mathbb{1}, i\gamma_5 \tau_a\}$, reads:

$$\Pi^{S/P}(\boldsymbol{p}, \omega_n) = \quad \bigcirc \quad = -T \sum_{m \in \mathbb{Z}} \int \frac{d^3q}{(2\pi)^3} \text{Tr}\left[\Gamma^{S/P} G_\beta^F(\boldsymbol{q}, \nu_m) \Gamma^{S/P} G_\beta^F(\boldsymbol{q}-\boldsymbol{p}, \nu_m - \omega_n)\right], \tag{3.91}$$

where the trace refers to all color, flavor and Dirac indices. In the calculation one has to take care of several minus signs: the fermion loop gives a global minus sign, in the pseudoscalar channel one has $i^2 = -1$ and additionally $\{\gamma_5, \gamma_\mu\} = 0$ and $\{\gamma_i, \gamma_j\} = -2\delta_{ij}$ for the Euclidean gamma matrices. One finds for the scalar and pseudoscalar channels:

$$\Pi^{S/P}(\boldsymbol{p}, \omega_n) = 4N_c|\widetilde{N}^{S/P}|T \sum_{m \in \mathbb{Z}} \int \frac{d^3q}{(2\pi)^3} \frac{\mp m^2 + \nu_m(\nu_m - \omega_n) + \boldsymbol{q}\cdot(\boldsymbol{q}-\boldsymbol{p})}{[\nu_m^2 + E_q^2][(\nu_m - \omega_n)^2 + E_\Delta^2]} =$$

$$= 2N_c|\widetilde{N}^{S/P}|T \sum_{m \in \mathbb{Z}} \int \frac{d^3q}{(2\pi)^3} \left[\frac{1}{\nu_m^2 + E_q^2} + \frac{1}{(\nu_m - \omega_n)^2 + E_\Delta^2}\right] + \tag{3.92}$$

$$+ 2N_c|\widetilde{N}^{S/P}|N^{S/P}T \sum_{m \in \mathbb{Z}} \int \frac{d^3q}{(2\pi)^3} \frac{1}{[\nu_m^2 + E_q^2][(\nu_m - \omega_n)^2 + E_\Delta^2]},$$

where we have introduced the notations

$$E_q^2 = \boldsymbol{q}^2 + m^2, \quad E_\Delta^2 = (\boldsymbol{q}-\boldsymbol{p})^2 + m^2, \quad \text{and} \quad E_\pm = E_q \pm E_\Delta, \quad N^P = -(\omega_n^2 + \boldsymbol{p}^2)\,. \tag{3.93}$$

In addition we use the prefactors $N^P = -(\omega_n^2 + \boldsymbol{p}^2)$, $N^S = N^P - 4m^2$, and $\widetilde{N}^S = -N_f$, $\widetilde{N}^P = 2$. Despite the fact that the integrals are divergent and need to be regularized we can formally shift the integration $\boldsymbol{q} \mapsto \boldsymbol{q} - \boldsymbol{p}$ and $\nu_m \mapsto \nu_m + \omega_n$ which suggests to define

$$I_1 = T \sum_{m \in \mathbb{Z}} \int \frac{d^3q}{(2\pi)^3} \frac{1}{\nu_m^2 + \boldsymbol{q}^2 + m^2} = \frac{1}{4\pi^2} \int_0^\Lambda dp \, \frac{p^2}{\sqrt{p^2 + m^2}} \left(1 - n_F^+(E_p) - n_F^-(E_p)\right). \tag{3.94}$$

The gap equation (in the two-flavor case) also involves the integral I_1, cf. Eq. (3.83):

$$m = m_0 - 2G\langle \bar{q}q \rangle = m_0 + 8GN_c m I_1\,. \tag{3.95}$$

Finally the polarization tensor becomes

$$\Pi^{S/P}(\boldsymbol{p}, \omega_n) = 4N_c|\widetilde{N}^{S/P}|I_1 + 2N_c|\widetilde{N}^{S/P}|N^{S/P}I_2(\boldsymbol{p}, \omega_n)\,, \tag{3.96}$$

where the momentum-dependent part can be expressed using $E_\pm = E_q \pm E_\Delta$:

$$
\begin{aligned}
I_2(\boldsymbol{p},\omega_n) &= T \sum_{m\in\mathbb{Z}} \int \frac{\mathrm{d}^3 q}{(2\pi)^3} \frac{1}{[\nu_m^2 + E_q^2][(\nu_m-\omega_n)^2 + E_\Delta^2]} = \\
&= \int \frac{\mathrm{d}^3 q}{(2\pi)^3} \frac{1}{4E_q E_\Delta} \Bigg[\frac{2E_+}{\omega_n^2 + E_+^2} + \\
&\quad + \frac{(\mathrm{i}\omega_n - E_+)\left(n_\mathrm{F}^-(E_q) + n_\mathrm{F}^+(E_\Delta)\right) - (\mathrm{i}\omega_n + E_+)\left(n_\mathrm{F}^+(E_q) + n_\mathrm{F}^-(E_\Delta)\right)}{\omega_n^2 + E_+^2} + \\
&\quad + \frac{(\mathrm{i}\omega_n + E_-)\left(n_\mathrm{F}^+(E_q) - n_\mathrm{F}^+(E_\Delta)\right) - (\mathrm{i}\omega_n - E_-)\left(n_\mathrm{F}^-(E_q) - n_\mathrm{F}^-(E_\Delta)\right)}{\omega_n^2 + E_-^2} \Bigg].
\end{aligned}
\tag{3.97}
$$

The Bethe-Salpeter equation (3.76) is given in a self-consistent form, $D_\mathrm{M} = G + G\Pi D_\mathrm{M}$. Its solution (3.72) has been already used in the derivation of the mesonic insertion $\Sigma_\mathrm{M}^{(1)}$ where this expression arose naturally from the derivative of a logarithm. We identify this term as the result of a geometric sum of coherent quark-antiquark loops:

$$
D_\mathrm{M}(\boldsymbol{p},\omega_n) = G + G\Pi D_\mathrm{M} = G\big(1 + \Pi G + (\Pi G)^2 + \ldots\big) = G\sum_{n=0}^{\infty}(\Pi G)^n = \frac{G}{1 - G\Pi(\boldsymbol{p},\omega_n)} . \tag{3.98}
$$

Poles in the meson propagator can appear only in Minkowskian metric, therefore we perform at this stage the analytical continuation $\mathrm{i}\omega_n \mapsto \omega + \mathrm{i}\epsilon$. Due to the fact that the polarization tensor is complex (when describing a resonance instead of a bounds state), we define the meson mass by the real pole position[28]:

$$
\mathrm{Re}\ D_\mathrm{M}^{-1}(\boldsymbol{0},-\mathrm{i}\omega)\big|_{\omega = m_\mathrm{M}} \stackrel{!}{=} 0 . \tag{3.99}
$$

If we concentrate now on the pseudoscalar (pionic) channel, we introduce $s = -N^\mathrm{P} = \omega_n^2 + \boldsymbol{p}^2 > 0$ and find with Eq. (3.98):

$$
D_\pi(\boldsymbol{p},\omega_n) = \frac{G}{\frac{m_0}{m} + 4GN_\mathrm{c}(\omega_n^2 + \boldsymbol{p}^2)I_2(\boldsymbol{p},\omega_n)} , \tag{3.100}
$$

from which the pion mass can be calculated by solving

$$
m_\pi^2 = m_\pi^2(\boldsymbol{0}) = \frac{m_0}{m}\frac{1}{4GN_\mathrm{c}\ \mathrm{Re}\ I_2(\boldsymbol{0},-\mathrm{i}m_\pi)} . \tag{3.101}
$$

If one considers the scalar channel instead, Π^S with $N^\mathrm{S} = N^\mathrm{P} - 4m^2$ in (3.92), the mass of the sigma boson can be extracted:

$$
m_\sigma^2(\boldsymbol{0}) = m_\sigma^2 = m_\pi^2 + 4m^2 \longrightarrow \approx m_\pi^2 \ \text{for large}\ T . \tag{3.102}
$$

We mention that there is in principle a momentum dependence of the pion mass, $m_\pi(\boldsymbol{p})$, due to the thermal environment and the implied loss of Lorentz invariance. Evaluating the pion mass in the rest frame, which is the frame of the heat bath, offers a consistent definition of a thermal pion mass. At least, this prescription approximates the interpretation of (pole) masses which are Lorentz-invariant quantities. In the reference frame of the heat bath, $\boldsymbol{p} = 0$, we have

$$
E_\Delta = E_q\ , \text{implying}\ \ E_+ = 2E_q\ , \ \ E_- = 0\ , \tag{3.103}
$$

---

[28] A further common strategy is to define the meson mass as maximizer of the spectral width. See also the discussion related to Fig. 3.7 where we compare these two approaches.

3. The Nambu–Jona-Lasinio model

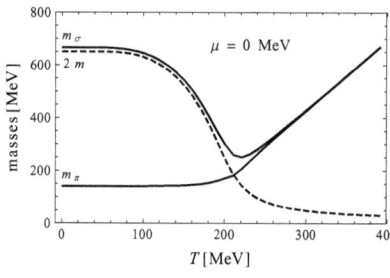

(a) Vanishing quark chemical potential

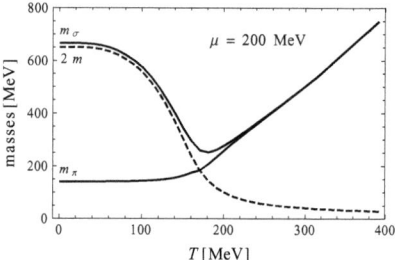

(b) Finite quark chemical potential $\mu = 200$ MeV

Figure 3.6.: Thermal masses of pions (lower solid line) and sigma boson (upper solid line). For high T the spontaneously broken chiral symmetry is restored and sigma and pion mass degenerate. We compare also to twice the constituent-quark mass (dashed line) which becomes small for high T. For temperatures above the Mott temperature, $T > T_\mathrm{M} \approx 212$ MeV, the pion becomes unstable and an on-shell decay channel into two constituent quarks opens.

leading to some significant simplifications for the polarization tensor in Eq. (3.96). Due to the now spherically symmetric integrand the angular integration is trivial and we arrive in Minkowski space at

$$I_2(\mathbf{0}, -\mathrm{i}\omega) = \frac{1}{2\pi^2} \int_0^\Lambda \mathrm{d}q \, \frac{q^2}{E_q} \frac{1 - n_\mathrm{F}^+(E_q) - n_\mathrm{F}^-(E_q)}{4E_q^2 - \omega^2} \,. \tag{3.104}$$

For all pairs (T, μ), this integral is real as long as the pole at $4E_q^2 = \omega^2$ is not hit and the mass ω describes a bound state. The resulting thermal meson masses are shown in Fig. 3.6: for small temperatures the (pseudo-)Goldstone-boson nature of the pion is indicated by a constantly small but non-vanishing mass. For high temperatures the pion looses its boundstate nature because an unphysical on-shell decay channel into a constituent quark-antiquark pair opens which defines the *Mott temperature* by $m_\pi(T_\mathrm{M}) = 2m(T_\mathrm{M})$. In this temperature region the pion mass is rising rapidly due to the restoration of (approximate) chiral symmetry. The pion becomes degenerate with the sigma boson which is not a Goldstone boson and features its minimal mass in the vicinity of the Mott temperature. For a non-vanishing quark chemical potential the temperature dependence of the meson masses remains qualitatively the same, but the Mott temperature is shifted to smaller values.

For resonances the integral $I_2(\mathbf{0}, \mathrm{i}\omega)$ is interpreted as principal value integral with additional imaginary part using Eq. (A.28). The poles are located at

$$q = q_0 = \sqrt{\frac{\omega^2}{4} - m^2} \,. \tag{3.105}$$

From this we evaluate the imaginary part of the polarization function:

$$\mathrm{Im}\, I_2(-\mathrm{i}\omega) = \frac{q_0}{8\pi\omega} \frac{\sinh\left(\frac{\omega}{2T}\right)}{\cosh\left(\frac{\mu}{T}\right) + \cosh\left(\frac{\omega}{2T}\right)} \Theta(\omega - 2m)\, \Theta(\Lambda - q_0) \,. \tag{3.106}$$

The first Θ-function accounts for the threshold of an on-shell decay of mesons into quark and antiquark. The second Θ-term is a pure cutoff effect and is irrelevant for energies below twice the cutoff, $\omega < 2\sqrt{\Lambda^2 + m^2}$ which always exceeds 1.3 GeV in the NJL model.

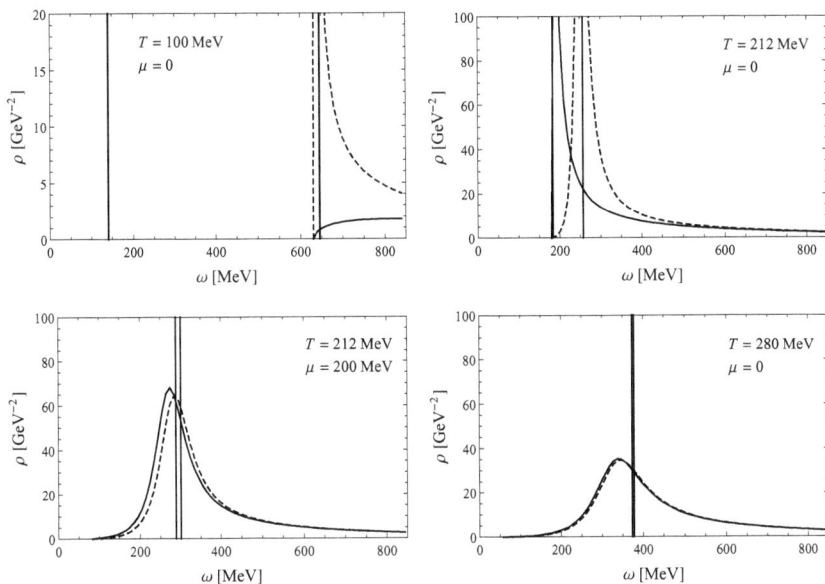

Figure 3.7.: Mesonic spectral function ρ for the pion (solid line) and sigma boson (dashed line). The two vertical solid lines are delta-function peaks located at the real poles determined from the condition (3.101).

In order to discuss in more detail the underlying physics of the (rescaled) meson propagators (3.98), we derive in a first step the corresponding spectral function $\rho(\omega)$. For propagators in Minkowski space it is defined by

$$G_{\mathrm{R/A}}(p_0, \boldsymbol{p}) = \int_{-\infty}^{\infty} \mathrm{d}\omega \, \frac{\rho(\omega, \boldsymbol{p})}{p_0 - \omega \pm \mathrm{i}\epsilon} \,, \tag{3.107}$$

with the "inverse" relation

$$\operatorname{Im} G_{\mathrm{R/A}}(p_0, \boldsymbol{p}) = \mp \pi \rho(p_0, \boldsymbol{p}) \,. \tag{3.108}$$

For example, the spectral function of a free bosonic propagator reads $\rho_0^{\mathrm{B}}(p^2) = Z \, \delta(p^2 - m_{\mathrm{B}}^2)$, if one takes the Feynman pole prescription ($+\mathrm{i}\epsilon$) into account. Thereby m_{B} denotes the mass of the boson and Z denotes its wave-function renormalization constant which defines in the context of mesons the quark-meson coupling as we will discuss in the next section. A negative spectral function arises if one considers instead the prescription ($-\mathrm{i}\epsilon$). Physical spectral functions satisfying causality constraints are positive. Then the advanced propagation of particles and retarded propagation of antiparticles is ensured. More details of the spectral function and the convention we use in this thesis can be found in Appendix A.3.

For an interacting theory, the spectral function contains generally contributions from the one-particle state, bound states and a continuum of two- and more-particle states. We find for the

3. The Nambu–Jona-Lasinio model

rescaled pion propagator $D_\pi(\mathbf{0}, -i\omega)$ from Eq. (3.100):

$$\rho_\pi(\omega) = g_{\pi qq}^2 \delta(\omega^2 - m_\pi^2) + \frac{1}{\pi} \frac{4G^2 N_c \omega^2 \operatorname{Im} I_2(\mathbf{0}, -i\omega)}{\left(\frac{m_0}{m} - 4GN_c \omega^2 \operatorname{Re} I_2(\mathbf{0}, -i\omega)\right)^2 + \left(4GN_c \omega^2 \operatorname{Im} I_2(\mathbf{0}, -i\omega)\right)^2} \,. \tag{3.109}$$

The spectral function for different pairs of (T, μ) is shown in Fig. 3.7. For small temperatures the isolated delta peak at $\omega = m_\pi$ displays the pion as (approximate) Goldstone boson of the spontaneous breakdown of chiral symmetry. The continuous part of ρ starts just for $\omega > 2m(T, \mu)$. The spectral function of the sigma boson peaks close to this threshold and exceeds significantly the spectral function of the pion. At the Mott temperature, the discrete and continuous spectrum are no longer separated and the pion propagator, D_π, has no longer a pole at a purely real value of m_π (cf. the mass definition in Eq. (3.99)). At even higher values of T or μ, the spectral functions show a maximum structure which can also be used to define the thermal meson masses. However, we also show in this region the real-pole positions indicated again by the solid vertical lines. As also seen in Fig. 3.6(b), they degenerate in the high-temperature limit.

3.5.2. Quark-meson coupling beyond the pole-mass approximation

The solution of the Bethe-Salpeter equation (3.100) in the pseudoscalar channel describes a rescaled pion propagator. In Minkowski metric it reads

$$D_\pi(\omega, \mathbf{p}) = \frac{G}{\frac{m_0}{m} - 4GN_c\left(\omega^2 - \mathbf{p}^2\right) I_2(\mathbf{p}, -i\omega)} \,. \tag{3.110}$$

where the pion mass is just determined by the solutions of Eq. (3.101). The fact that I_2 is energy and momentum dependent implies that the standard pole-mass approximation,

$$\widetilde{D}_\pi(\omega, \mathbf{p}) = \frac{g_{\pi qq,\text{pole}}^2}{\omega^2 - \mathbf{p}^2 - m_\pi^2 + i\epsilon} \,, \tag{3.111}$$

with a momentum-independent constant $g_{\pi qq,\text{pole}}$ does not reproduce the $(\omega, \mathbf{p})$ dependence of the full meson propagators. The full quark-pion coupling can be obtained from

$$g_{\pi qq}(\omega, \mathbf{p})^2 = \left(\omega^2 - \mathbf{p}^2 - m_\pi^2(\mathbf{p})\right) D_\pi(\omega, \mathbf{p}) \,, \tag{3.112}$$

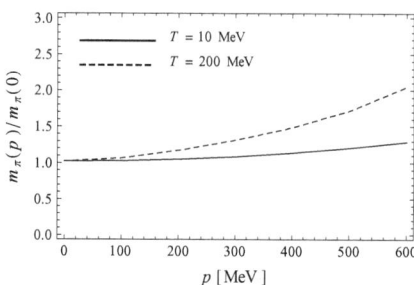

Figure 3.8.: Momentum dependence of the pion mass at different temperatures

3.5. Bethe-Salpeter equation and thermal meson masses

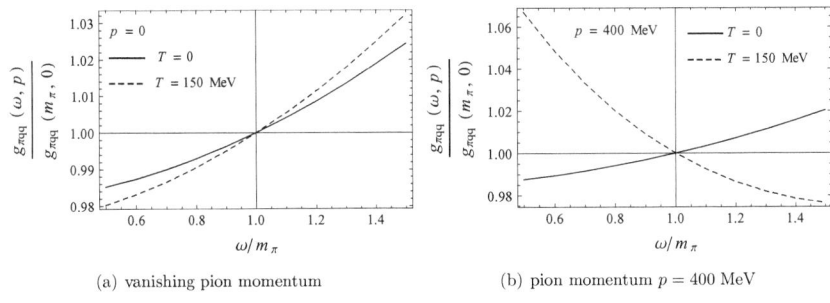

(a) vanishing pion momentum

(b) pion momentum $p = 400$ MeV

Figure 3.9.: Corrections to the quark-pion coupling in a 50% range around the pole position

with a momentum-dependent pion mass defined by

$$\operatorname{Re} D_\pi^{-1}(m_\pi(\boldsymbol{p}), \boldsymbol{p}) = 0 \ . \tag{3.113}$$

Here we must not evaluate the pion mass just using the conditional equation (3.101) but we need to incorporate the full momentum dependence coming from Eq. (3.100). The results for the momentum dependent pion mass are shown in Fig. 3.8: for small and large temperatures the pion becomes more massive when it carries additional momentum. This qualitative behavior of $m_\pi(p)$ is consistent with the fact that the constituent quark mass $m(p)$ decreases as function of momentum meaning that at high momenta chiral symmetry is restored where pions can no longer be interpreted as Goldstone bosons of spontaneous chiral symmetry breaking.

Usually, the quark-pion coupling is defined as residue of the full pion propagator at vanishing momentum [HK87]:

$$g_{\pi qq}^{-2} = \frac{1}{\operatorname{Res} D_\pi} = \left. \frac{\mathrm{d}(D_\pi(\boldsymbol{0}, -i\omega))^{-1}}{\mathrm{d}\omega^2} \right|_{\omega^2 = m_\pi^2} . \tag{3.114}$$

In pole-mass approximation one finds immediately

$$g_{\pi qq,\text{pole}} = \frac{1}{\sqrt{4N_c I_2(\boldsymbol{0}, -im_\pi)}} \ . \tag{3.115}$$

Taking the remaining energy dependence of $I_2(\boldsymbol{0}, -i\omega)$ into account, we find

$$\begin{aligned} g_{\pi qq}^{-2}(\omega, \boldsymbol{0}) &= \frac{1}{g_{\pi qq,\text{pole}}^2} \left(1 + m_\pi^2 \frac{1}{I_2(\boldsymbol{0}, -im_\pi)} \left. \frac{\mathrm{d} I_2(\boldsymbol{0}, -i\omega)}{\mathrm{d}\omega^2} \right|_{\omega^2 = m_\pi^2} \right) \\ &= 4N_c \left(I_2(\boldsymbol{0}, -im_\pi) + m_\pi^2 \left. \frac{\mathrm{d} I_2(\boldsymbol{0}, -i\omega)}{\mathrm{d}\omega^2} \right|_{\omega^2 = m_\pi^2} \right) . \end{aligned} \tag{3.116}$$

Taking the derivative of the principal-value integral (3.104) leads to some highly singular expression near to pole $q = \pm q_0$. Exactly at threshold of an on-shell decay of the pion into two quarks (with zero momenta) the derivative of I_2 becomes divergent. In order to remove the second-order pole in the derivative, we integrate by parts. The surface term vanishes for all T and μ when the

3. The Nambu–Jona-Lasinio model

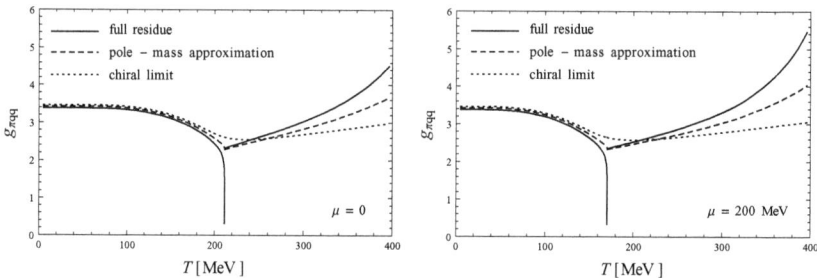

Figure 3.10.: Temperature dependence of the quark-meson coupling for different scenarios: full residue of the pion propagator but set $\boldsymbol{p} = 0$ (solid line, Eq. (3.116)) in comparison to the pole-mass approximation (dashed line, Eq. (3.115)) and in the chiral limit (dotted line). The qualitative dependence on μ is much less pronounced than its pure temperature dependence.

cutoff $\Lambda \to \infty$. The remaining expression needs to be treated again as principal-value integral:

$$\begin{aligned}\frac{\mathrm{d} I_2(\mathbf{0}, -\mathrm{i}\omega)}{\mathrm{d}\omega^2}\bigg|_{\omega^2 = m_\pi^2} &= \frac{1}{2\pi^2} \int_0^\Lambda \mathrm{d}q\, \frac{q}{(4E_q^2 - \omega^2)^2} \frac{q}{E_q} \left(1 - n_\mathrm{F}^+(E_q) - n_\mathrm{F}^-(E_q)\right) = \\ &= \frac{1}{2\pi^2} \int_0^\Lambda \mathrm{d}q\, \frac{1}{8(4E_q^2 - \omega^2)} \cdot \left[\frac{1}{E_q}\left(1 - \frac{q^2}{E_q^2}\right)\left(1 - n_\mathrm{F}^+(E_q) - n_\mathrm{F}^-(E_q)\right) \right. \\ &\quad \left. + \frac{\beta q^2}{E_q^2}\left[n_\mathrm{F}^+(E_q)\bigl(1 - n_\mathrm{F}^+(E_q)\bigr) + n_\mathrm{F}^-(E_q)\bigl(1 - n_\mathrm{F}^-(E_q)\bigr)\right]\right]. \end{aligned}$$
(3.117)

The pole structure of this result is the same as found for I_2 in Eq. (3.104). As we discuss in detail for a set of prototype functions in App. A.4, the integration by parts yields a principal-value integral which diverges for the pole approaching the interval of integration $[0, \Lambda]$.

In Fig. 3.9 we compare the two approaches for calculating the quark-pion coupling. When staying in a 50% interval around the real pion pole, $0.5 m_\pi < \omega < 1.5 m_\pi$, we find that the usual treatment is a good approximation. It is interesting to note that at small (i.e. vanishing) momenta at $T = 150$ MeV the approximate quark-pion coupling overestimates the actual coupling when evaluating at energies smaller than the pion mass. For high momenta but at the same temperature the coupling is underestimated by up to 6% which is still a reasonable approximation. Sizable deviations occur only at very high temperatures and unreasonably far away from the actual pole mass. However, apart from such extreme values, we conclude that the pole-approximation of the quark-pion coupling is acceptable and corrections beyond Eq. (3.116) are within a few percent. Therefore, we do not take any momentum dependence into account apart from the correction in Eq. eqrefQuarkMesonCouplingDerivateResult. This leads to a simplified treatment of the mesonic fluctuations in Chapter 5 since the quark-meson coupling is just a constant and therefore not affected by the momentum integration.

In Fig. 3.10 we show the results for the quark-meson coupling as function of the temperature for three different cases: the solid line refers to the calculation of the full residue, where also the derivative of I_2 is taken into account. We compare this to the cases where the pion mass is fixed either by the the physical quark masses or in the chiral limit: the solid line features a jump starting from a vanishing coupling constant. Fixing the pion mass first leads to continuous coupling constants with a kink (for physical quark masses) or a very smooth temperature dependence (in the chiral limit). Our results including the derivative terms agree well with the steep decrease of $g_{\pi qq}$ for T approaching the Mott temperature [HK87]. For all three scenarios the low-temperature behavior is almost the same: below $T = 100$ MeV, the quark-meson coupling is constant, with slightly different values. This behavior is in agreement with earlier results from chiral effective field theory [EK94], where perturbative calculations show that for small temperatures $g_{\pi qq}$ is independent of T up to quadratic order $\mathcal{O}(T^2)$.

In summary we have found that the momentum-dependence of the quark-meson coupling is not crucial and can be neglected, allowing for an efficient numerical treatment. The remaining energy-dependence of I_2 is important and affects the quark-meson coupling qualitatively as compared to the pole-mass approximation. In the chiral limit the two approaches coincide, as seen from Eq. (3.116). In the Nambu-Goldstone phase the pion becomes massless and one finds:

$$\lim_{m_0 \to 0} g_{\pi qq} = g_{\pi qq,\text{pole}} = \frac{Gmm_\pi^2}{m_0} . \tag{3.118}$$

Inserting the pion mass (3.101) into the right-hand side and inspecting Eqs. (3.115) and (3.116), one can prove this identity immediately.

3.6. Pion decay constant and low-energy theorems

In this section we demonstrate that important low-energy theorems known from QCD as the Gell-Mann-Oakes-Renner (GOR) and the Goldberger-Treiman (GT) relations [LK96, CL06] are fulfilled in the NJL model. In Section 2.1.2 these relations have been derived directly from QCD, see the GOR relation in Eq. (2.32). The pion decay constant is defined by the transition amplitude, Eq. (2.24), where the pion is annihilated through the axialvector current into the vacuum. Diagrammatically we find in the vacuum, i.e. for $T = 0$ and $\mu = 0$:

$$ip^\mu f_\pi \delta_{ab} = \quad \cdots \cdots \cdots \quad =$$

$$= -\int \frac{d^4q}{(2\pi)^4} \text{Tr} \left[\gamma^\mu \gamma_5 \frac{\lambda_a}{2} G_M^F(q) \left(-ig_{\pi qq}\gamma_5 \lambda_b\right) G_M^F(q+p) \right] =$$

$$= \delta_{ab} N_c \, ig_{\pi qq} \int \frac{d^4q}{(2\pi)^4} \frac{\text{tr}\left[\gamma^\mu \gamma_5 (\slashed{q}+m)\gamma_5(\slashed{q}+\slashed{p}+m)\right]}{[(q^2-m^2)][(q+p)^2-m^2]} , \tag{3.119}$$

where the color and flavor traces result in a global factor of N_c and the Kronecker delta due to $\text{tr}(\lambda_a \lambda_b) = 2\delta_{ab}$ for the fundamental representation (cf. Table A.1 in the Appendix). The remaining trace in the integrand simplifies since for both terms, m^0 and m^2, an odd number of Dirac matrices occur, so only the linear term in m contributes: $\text{Tr}\left[\gamma^\mu \gamma_5 \left(m\gamma_5(\slashed{p}+\slashed{q}) + m\slashed{q}\gamma_5)\right)\right] = 4mp^\mu$. From this we find

$$ip^\mu f_\pi \delta_{ab} = \delta_{ab} N_c \, ig_{\pi qq} \cdot 4mp^\mu \int \frac{d^4q}{(2\pi)^4} \frac{1}{[(q^2-m^2)][(q+p)^2-m^2]} , \tag{3.120}$$

where the remaining integral is just the Minkowski version of $I_2(\boldsymbol{q}, \omega_m)$ defined in Eq. (3.97). For the thermal pion-decay constant we find

$$f_\pi(\boldsymbol{p}, \omega_n) = 4m N_c g_{\pi qq} I_2(\boldsymbol{p}, \omega_n) . \tag{3.121}$$

3. The Nambu–Jona-Lasinio model

From this both low-energy theorems follow, starting with the GT relation:

$$f_\pi g_{\pi qq} = 4mN_c g_{\pi qq}^2 I_2 = m + \mathcal{O}(m_0) \,, \tag{3.122}$$

where we have expanded the quark-pion coupling (3.116) around the chiral limit $m_0 = 0$ up to first order in $\mathcal{O}(m_0) = \mathcal{O}(m_\pi^2)$ and evaluated at $\boldsymbol{p} = 0$ and $\omega_n \mapsto -im_\pi$. Also the GOR relation follows immediately from Eqs. (3.101) and (3.120):

$$m_\pi^2 f_\pi^2 = \frac{m_0}{m} \frac{1}{4GN_c I_2} \cdot 16m^2 N_c^2 g_{\pi qq}^2 I_2^2 = m_0 \cdot \frac{4N_c I_2 g_{\pi qq}^2 m}{G} \,. \tag{3.123}$$

Expanding again the quark-pion coupling around the chiral limit, $g_{\pi qq}^2 = (4N_c I_2)^{-1} + \mathcal{O}(m_0)$ and inspecting the gap equation (3.83) up to the same order, $m = -G\langle\bar{\psi}\psi\rangle + \mathcal{O}(m_0)$, the GOR relation is found to be valid in the NJL model as well:

$$m_\pi^2 f_\pi^2 = -m_0 \langle\bar{\psi}\psi\rangle + \mathcal{O}(m_0^2) \,. \tag{3.124}$$

In conclusion the important low-energies theorems of QCD hold in the NJL model as they should. Since this model is based on the symmetry patterns of QCD, proving their validity is a crucial crosscheck for physical consistency.

3.7. Attractive diquark channels

The general discussion of QCD color-color currents, cf. Table 3.2, has shown that both the color-singlet channel and the totally-antisymmetry channel contribute for the physical number of colors, $N_c = 3$, with similar importance. The first one refers to quark-antiquark correlations with channel weight $-8/3$, whereas the $\bar{r}_a$ channel refers to quark-quark correlations with channel weight $-4/3$. Diquarks are colored objects, therefore, due to color confinement, they do not exist in the physical spectrum. Within the NJL model that does not describe confinement, diquarks should be considered as possible degrees of freedom in addition to the mesonic modes.

In Section 3.1 we have introduced the Fierz transformation which also generates quark-quark interactions [VW91]:

$$\begin{aligned}\mathcal{L}_{qq} &= \sum_D H_D (\bar{\psi} \Gamma^{(D)} C \bar{\psi}^T)(\psi^T C \Gamma^{(D)} \psi) = \\ &= H_S \sum_{f_A, c_A} \left[\left(\bar{\psi} i\gamma_5 t^c \lambda_A C \bar{\psi}^T\right) \left(\psi^T C i\gamma_5 t^c \lambda_A \psi\right) + \ldots \right] \ldots \,, \end{aligned} \tag{3.125}$$

where in the first set of dots the pseudoscalar interaction (without the $i\gamma_5$ term) is included[29], and the second set of dots contains vector and axialvector contributions and those from the totally symmetric channel r_s. We have denoted the charge-conjugation operator by $C = i\gamma_0\gamma_2$ with $C^{-1} = C^\dagger = -C$. In $\Gamma^{(D)}$ Dirac, flavor and color structures are encoded, thus for the quark-antiquark interactions, in principle, also color-adjoint (color-octet) terms might appear. Masses of such objects appear to be larger than 2 GeV $\approx 3\Lambda$ and are therefore suppressed compared to the typical scales in our NJL model [Tha06]. Also from the viewpoint of large-N_c analysis such objects are suppressed since the color-adjoint channel is repulsive and becomes weak for $N_c \to \infty$. The totally symmetric channel r_s is repulsive for all values of N_c and its strength is bounded from above, therefore it is not considered in our calculations.

---

[29]Note that due to the appearance of the charge-conjugation operator C the behavior of the diquark currents under parity transformations is opposite to the naively expected ones.

3.7. Attractive diquark channels

In contrast to the quark-antiquark channel, the Pauli principle constrains the quark-quark channels: the requirement of a totally antisymmetric wavefunction for the considered two-quark system implies that the totally anti-symmetric color channel $\bar{r}_a$ must also have an anti-symmetric flavor part. In the totally symmetric color channel r_s only the symmetric flavor part of the flavor space is relevant. In conclusion, for diquarks the flavor and color spaces are no longer independent but are entangled via the Pauli principle[30]. Every single Gell-Mann matrix (cf. Appendix A.1) is either symmetric, $\lambda_i^T = \lambda_i$, or antisymmetric, $\lambda_i^T = -\lambda_i$, no mixed symmetries appear. In Section 3.1 we have already derived the number of symmetric and antisymmetric generators of $SU(N_c)$:

$$\dim(r_s) = \frac{N_c(N_c+1)}{2}, \quad \dim(\bar{r}_a) = \frac{N_c(N_c-1)}{2}. \tag{3.126}$$

In general we find

$$\sum_{i=1}^{N_c^2-1} \lambda_{ab}^i \lambda_{dc}^i = \frac{N_c-1}{N_c} \sum_{r_s} \lambda_{ad}^s \lambda_{cb}^s - \frac{N_c+1}{N_c} \sum_{\bar{r}_a} \lambda_{ad}^a \lambda_{cb}^a, \tag{3.127}$$

which evaluates for SU(3) to:

$$\sum_{i=1}^{8} \lambda_{ab}^i \lambda_{dc}^i = \frac{2}{3} \sum_{r_s} \lambda_{ad}^s \lambda_{cb}^s - \frac{4}{3} \sum_{\bar{r}_a} \lambda_{ad}^a \lambda_{cb}^a. \tag{3.128}$$

From a Fierz-transformed color-color current in Eq. (3.1), the couplings G_S and H_S are fixed:

$$G_S = \frac{N_c^2-1}{N_c^2} G_c, \quad H_S = \frac{N_c+1}{2N_c} G_c. \tag{3.129}$$

From this, also their ratio is fixed,

$$\frac{H_S}{G_S} = \frac{N_c}{2(N_c-1)} \to \frac{1}{2} \quad \text{as} \quad (N_c \to \infty). \tag{3.130}$$

We reproduce the well-known values for the physical case $N_c = 3$:

$$G_S = \frac{8}{9} G_c, \quad H_S = \frac{2}{3} G_c \quad \Rightarrow \quad \frac{H_S}{G_S} = \frac{3}{4}. \tag{3.131}$$

In the literature the chosen values are around this Fierz-induced value for the ratio H_S/G_S. For instance, in [VW91] one finds

$$H_S = 2G_V - \frac{1}{3}(\delta G_V + \delta G_A) = 0.97 \, G_S, \tag{3.132}$$

with the values for $G_V, \delta G_V, \delta G_A$ used in the reference. For our calculation we chose an even larger value $H_S = 1.2 G_S$ as discussed below.

The Bethe-Salpeter equation for diquarks is derived analogously to the quark-antiquark channel via iterations of quark-quark loops in an infinite ladder. This is again just the random-phase approximation and follows from a large-N_c expansion at next-to-leading order. The diquark

---

[30] In the literature the phase where diquark condensates are dominant is called "color-flavor locked" (CFL) phase in order to express this entanglement.

3. The Nambu–Jona-Lasinio model

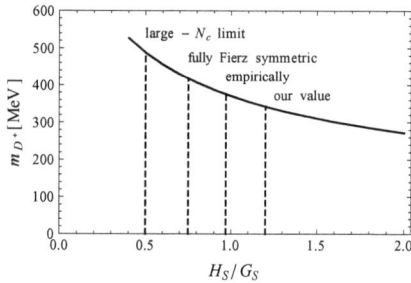

Figure 3.11.: Vacuum value of the scalar diquark mass for different ratios if the coupling H_S/G_S for fixed G_S. With the vertical lines we denote specific ratios as they can be derived in the large-N_c limit or assuming a fully Fierz-invariant NJL Lagrangian (cf. the discussion in the text).

loop reads (cf. the quark-antiquark loop Eq. (3.91) for the mesonic case):

$$\Pi_{\rm D}^{ij}(\omega_n, \bm{p}) = -T \sum_{m\in\mathbb{Z}} \int \frac{d^3q}{(2\pi)^3} \,{\rm Tr}\left[\Gamma^i C G_\beta^{\rm F}(\bm{q},\nu_m) \Gamma^j C^{-1} \left(G_\beta^{\rm F}(\bm{p}-\bm{q},\nu_n-\nu_m)\right)^{\rm T}\right], \qquad (3.133)$$

where the trace covers color, flavor and Dirac space. The color and flavor structures are already fixed by the Pauli principle to be antisymmetric. The underlying Lagrangian (3.125) offers only two channels, $\Gamma^{\rm P} = {\rm i}\lambda^A t^c$ and $\Gamma^{\rm S} = \gamma_5 \lambda^A t^c$. We mention again that the presence of γ_5 leads in context of diquark to *scalar* currents.

Using $[C, \gamma_5] = 0$ and $[C, \gamma_\mu] = 0$, one derives in the scalar channel for the Dirac part of the loop:

$$\Pi_{\rm D}^{\rm SS}(\bm{p},\omega_n) = {\rm tr}_{\rm D}\left[\gamma_5 G_\beta^{\rm F}(\bm{q},\nu_m)\gamma_5 G_\beta^{\rm F}(\bm{q}-\bm{p},\nu_m-\nu_n)\right], \qquad (3.134)$$

where the trace to Dirac space only. Note the relative minus sign in the second quark propagator: $\bm{p}-\bm{q} \mapsto \bm{q}-\bm{p}$. This is exactly the same structure as in the mesonic case. The only difference are additional non-trivial factors from color and flavor space:

$$\operatorname{tr}(\lambda_a^{\rm r} \lambda_b^{\rm r}) = \delta_{ab}\xi(r)\dim(r) . \qquad (3.135)$$

We need to sum over all antisymmetric generators in color space,

$$\operatorname{tr}_{C_A}(t^c t^c) = \frac{1}{4}\sum_{C_a} \xi(F)\dim(F) = \frac{1}{4}\dim(\bar{r}_{\rm a}) \cdot \frac{2}{N_c} \cdot N_c = \frac{N_c(N_c-1)}{4}, \qquad (3.136)$$

and flavor space,

$$\operatorname{tr}_{F_A}(\lambda_A \lambda_A) = N_f(N_f - 1) . \qquad (3.137)$$

For $N_f = 2$ this is just the same numerical factor $\operatorname{tr}_{F_A} = 2$ as in the mesonic sector, but the underlying physics is quite different! In color space the trivial factor N_c is substituted by Eq. (3.136). We mention that for the two-flavor case with $N_c = 5$ the numerical factors in the quark-quark and quark-antiquark channels coincide.

Solving the Bethe-Salpeter equation for the scalar and pseudoscalar diquark channels yields the vacuum diquark masses as shown in Fig. 3.11. They depend strongly on the coupling H_S

3.7. Attractive diquark channels

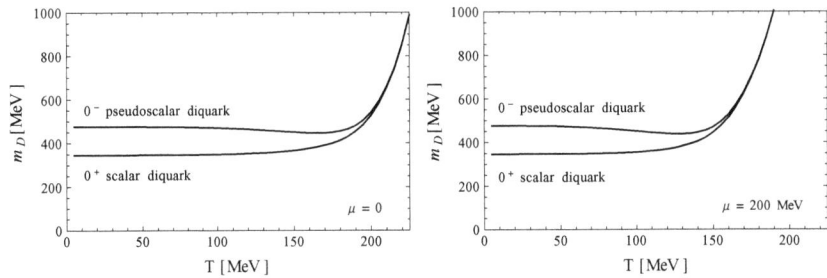

Figure 3.12.: Temperature dependence of diquark masses in the scalar (0^+) and pseudoscalar (0^-) channel for $\mu = 0$ and $\mu = 200$ MeV

and become smaller when the coupling increases. The empirical value for the diquark coupling suggests a diquark mass of $m_{D^+} \approx 400$ MeV, which is shifted to even higher values when assuming a fully Fierz-invariant NJL Lagrangian or when taking the value from the large-N_c limit. For calculating the thermal dependences of the diquark masses we have chosen the coupling to be $H_S = 1.2 G_S$. Results for the thermal diquark masses as shown in Fig. 3.12. Since both the mesonic and diquark spectrum is determined from a Bethe-Salpeter equation with similar kernels, also their thermal dependence is comparable, cf. Fig. 3.6. But there are also differences: in the mesonic spectrum the pseudoscalar meson (pion) is lighter than the scalar (sigma) meson. As already stated before, in the diquark spectrum this order is reversed and the scalar diquark (0^+) is lighter than the pseudoscalar diquark (0^-). For our purposes the temperature range $T > 200$ MeV is most important. There, the diquark masses are much larger than the meson masses as pointed out again in Table 3.6. We conclude that diquarks do not provide sizable fluctuations contributing to the shear viscosity in the end. Our results for m_D and the values in [VW91] are smaller than those from QCD Dyson-Schwinger equations (in rainbow-ladder truncation), where the diquark masses are at least 800 MeV [Mar02] or even 1.4 GeV [WLC+13]. Also (quenched) lattice simulations provide large diquark masses above 690 MeV [HKLW98].

T [MeV]	m_π [GeV]	m_σ [GeV]	m_{D^+} [GeV]
200	0.17	0.30	0.53
240	0.26	0.28	1.87

Table 3.6.: Comparison of the meson masses to the lightest (scalar) diquark with $J^P = 0^+$ at vanishing quark chemical potential

4. Microscopic theory of the shear viscosity

> *"Ich behaupte aber, daß in jeder besonderen Naturlehre nur so viel eigentliche Wissenschaft angetroffen werden könne, als darin Mathematik anzutreffen ist."*
> [Kan86]
>
> Immanuel Kant

In this chapter we present the Kubo formalism for the shear viscosity in the NJL model. Starting from the general result discussed in Section 2.3 we derive that the (static) shear viscosity η can be interpreted as the slope of some retarded correlator in its low-frequency limit. The large-N_c counting scheme is applied to the Kubo formalism in order to find leading-order contributions to η. The main purpose of this chapter is the investigation of inherent physical and numerical properties encoded in the structure of the Kubo formula for η. We introduce a convenient numerical approximation scheme and discuss the very strong cutoff dependence of the shear viscosity. Purely parametrically we also investigate how the shape of the (momentum dependent) spectral width $\Gamma(p)$ influences the overall thermal behavior of $\eta[\Gamma(p)]$. This is done in order to explore the inherent thermal effects on η before calculating a physical quark spectral width in Chapter 5. Selected results of the present chapter have been published previously in [LW13].

4.1. Shear viscosity from Kubo formalism

In the Kubo formalism transport coefficients are related to four-point functions of the energy-momentum tensor in Matsubara space. Using the NJL model as representative of a "typical" fermionic theory, we refer back to the Kubo formula for the shear viscosity in Eq. (2.71). In general the shear viscosity is a field but for calculations it is evaluated at the origin of Minkowski space, $x = (t, \boldsymbol{x}) = (0, \boldsymbol{0})$, assuming an infinite homogeneous medium close to thermal equilibrium:

$$\eta(\omega) = \frac{\beta}{10} \int_0^\infty dt\, e^{i\omega t} \int d^3x\, (\pi_{\mu\nu}(t,\boldsymbol{x}), \pi^{\mu\nu}(0,\boldsymbol{0})), \tag{4.1}$$

where the viscous-stress tensor $\pi_{\mu\nu}$ denotes the traceless part of the energy momentum tensor, defined in Eq. (2.65). The correlator in the integrand of the Kubo formula has been defined in Eq. (2.72) expressing the linear response of the energy-momentum tensor to the dissipative force, cf. the general discussion in Section 2.3. In the literature one finds alternative representations of the Kubo formula (4.1), e.g.

$$\begin{aligned}\eta(\omega) &= \beta \int_0^\infty dt\, e^{i\omega t} \int d^3x\, (T_{21}(t,\boldsymbol{x}), T_{21}(0,\boldsymbol{0})) = \\ &= \frac{\beta}{15} \int_0^\infty dt\, e^{i\omega t} \int d^3x\, (T_{\mu\nu}(t,\boldsymbol{x}), T^{\mu\nu}(0,\boldsymbol{0})),\end{aligned} \tag{4.2}$$

where in the first line only one component of the energy-momentum tensor contributes whereas in the second line summation takes place. The factor 15 is due the integration in an isotropic

4. Microscopic theory of the shear viscosity

medium ($i, j \in \{1, 2, 3\}$ with $i \neq j$):

$$\int d\Omega\, x_i^2 x_j^2 f(x^2) = \frac{1}{15} \int d\Omega\, x^4 f(x^2)\,. \tag{4.3}$$

With $\pi_{\mu\nu}\pi^{\mu\nu} = \frac{2}{3} T_{\mu\nu} T^{\mu\nu}$ one can see that all three representations of the shear viscosity $\eta(\omega)$ coincide.

Since the energy-momentum tensor $T^{\mu\nu}$ and therefore also the viscous-stress tensor $\pi^{\mu\nu}$ are real quantities[31], so that:

$$\eta(\omega)^* = \eta(-\omega) \in \mathbb{C} \quad \Rightarrow \quad \eta = \lim_{\omega \to 0} \eta(\omega) \in \mathbb{R}\,. \tag{4.4}$$

Taking the low-frequency limit of the shear viscosity (field) $\eta(\omega; x)$ introduces the so-called *static shear viscosity* η which is indeed a (positive) real number. When talking from now on about shear viscosity we will always refer to the static η. We will now rewrite $\eta(\omega)$ in more suitable form and introduce the retarded correlator

$$\Pi^R(\omega) = -\mathrm{i} \int d^3x \int_0^\infty dt\, e^{\mathrm{i}\omega t}\, \langle [T_{21}(t, \boldsymbol{x}), T_{21}(0, \boldsymbol{0})]\rangle_0\,, \tag{4.5}$$

where the term "retarded" is due to the fact that the first argument in the commutator, $T_{21}(t, \cdot)$, is evaluated only at later times $t \in (0, \infty)$ compared to the second argument, $T_{21}(0, \cdot)$. Putting the equilibrium fluctuations of the energy-momentum tensor to zero, $\langle T_{\mu\nu}(x)\rangle_0 = 0$, we can calculate

$$\begin{aligned}
\eta(\omega) &= \beta \int_0^\infty dt\, e^{\mathrm{i}\omega t} \int d^3x\, (T_{21}(t, \boldsymbol{x}), T_{21}(0)) = \\
&= \int_0^\infty dt\, e^{\mathrm{i}\omega t} \int d^3x \int_0^\beta d\xi\, \langle T_{21}(t, \boldsymbol{x})\, e^{-\xi H}\, T_{21}(0)\, e^{\xi H}\rangle_0 = \\
&= \int_0^\infty dt\, e^{\mathrm{i}\omega t} \int d^3x \int_0^\beta d\xi\, \langle T_{21}(t - \mathrm{i}\xi, \boldsymbol{x}) T_{21}(0)\rangle_0 = \\
&= \mathrm{i} \int d^3x \int_0^\infty dt\, e^{\mathrm{i}\omega t}\, \langle \big(K_{21}(t - \mathrm{i}\beta, \boldsymbol{x}) - K_{21}(t, \boldsymbol{x})\big) T_{21}(0)\rangle_0\,.
\end{aligned} \tag{4.6}$$

In the last line we have used $d\xi = \mathrm{i}\, dy$ for $y = t - \mathrm{i}\xi$ and have introduced a "potential" for the energy-momentum tensor by

$$K_{\mu\nu}(t, \boldsymbol{x}) = \int dt\, T_{\mu\nu}(t, \boldsymbol{x})\,, \tag{4.7}$$

where we can impose without loss of generality that $K_{\mu\nu}(\infty, \boldsymbol{x}) = 0$. Using this boundary condition we now integrate by parts and arrive at

$$\begin{aligned}
\eta(\omega) &= \frac{\mathrm{i}}{\omega} \int d^3x \int_0^\infty dt\, e^{\mathrm{i}\omega t} \langle \mathrm{i}\big(T_{21}(t - \mathrm{i}\beta, \boldsymbol{x}) - T_{21}(t, \boldsymbol{x})\big) T_{21}(0)\rangle_0 + \\
&\quad + \frac{\mathrm{i}}{\omega} \int d^3x\, \langle \mathrm{i}\big(K_{21}(-\mathrm{i}\beta, \boldsymbol{x}) - K_{21}(0, \boldsymbol{x})\big) T_{21}(0)\rangle_0\,.
\end{aligned} \tag{4.8}$$

Using the identity

$$\langle X(t) Y(t' + \mathrm{i}\beta)\rangle_0 = \langle Y(t') X(t)\rangle_0\,, \tag{4.9}$$

---

[31] To be more precise they are quantum operators, but each component is a real-valued function which can be seen by checking $T_{\mu\nu}^* = T_{\mu\nu}$ directly from its field-theoretical definition in Eq. (4.14).

4.1. Shear viscosity from Kubo formalism

we get for the first line of Eq. (4.8):

$$\int d^3x \int_0^\infty dt\, e^{i\omega t} \langle i(T_{21}(t-i\beta,\boldsymbol{x}) - T_{21}(t,\boldsymbol{x}))T_{21}(0)\rangle_0 =$$
$$= -i \int d^3x \int_0^\infty dt\, e^{i\omega t} \langle [T_{21}(t,\boldsymbol{x}), T_{21}(0)]\rangle_0 = \Pi^{\mathrm{R}}(\omega) \,. \tag{4.10}$$

For the second line of Eq. (4.8) we use again $K_{\mu\nu}(\infty,\boldsymbol{x}) = 0$ and calculate in a similar way:

$$\int d^3x \langle i(K_{21}(-i\beta,\boldsymbol{x}) - K_{21}(0,\boldsymbol{x}))T_{21}(0)\rangle_0 =$$
$$= i \int d^3x \, \langle [T_{21}(0), K_{21}(0,\boldsymbol{x})]\rangle_0 = -\Pi^{\mathrm{R}}(0) \,. \tag{4.11}$$

In the end we arrive at the following simple expression for the shear viscosity:

$$\eta = \lim_{\omega\to 0} \eta(\omega) = \lim_{\omega\to 0} \frac{i}{\omega}\left(\Pi^{\mathrm{R}}(\omega) - \Pi^{\mathrm{R}}(0)\right) \,. \tag{4.12}$$

Since we pulled out a prefactor i/ω, also the retarded correlator Π^{R} inherits the same property as $\eta(\omega)$ when applying the complex conjugation: $\Pi^{\mathrm{R}}(\omega)^* = \Pi^{\mathrm{R}}(-\omega)$. This implies an even real part and an odd imaginary part of the retarded correlator. We arrive at the following representation of the shear viscosity:

$$\eta = i\left.\frac{d}{d\omega}\Pi^{\mathrm{R}}(\omega)\right|_{\omega=0} = -\left.\frac{d}{d\omega}\mathrm{Im}\,\Pi^{\mathrm{R}}(\omega)\right|_{\omega=0} \,. \tag{4.13}$$

Using now the NJL Lagrangian $\mathcal{L}_{\mathrm{NJL}} = \mathcal{L}_{\mathrm{kin}} + \mathcal{L}_{\mathrm{int}}$ in Eq. (3.31), the imaginary part of the retarded correlator $\Pi^{\mathrm{R}}(\omega)$ involves the kinetic term $\mathcal{L}_{\mathrm{kin}}$ only. Apart from the cutoff dependence, $G(p) = G\Theta(\Lambda - p)$, all NJL couplings are independent of the momentum. We find:

$$T_{\mu\nu} = \frac{\partial \mathcal{L}_{\mathrm{NJL}}}{\partial(\partial^\mu \psi)} \partial_\nu \psi - g_{\mu\nu}\mathcal{L}_{\mathrm{NJL}} = i\bar\psi\gamma_\mu \partial_\nu \psi - g_{\mu\nu}\mathcal{L}_{\mathrm{NJL}} \,. \tag{4.14}$$

The relevant parts for the shear viscosity are off-diagonal elements only, hence the second term which carries the interaction does not contribute to η, cf. Eq. (4.2). In order to evaluate the retarded correlator $\Pi^{\mathrm{R}}(\omega)$ defined in Eq. (4.5) we switch to Matsubara space and calculate first

$$\Pi(\omega_n) = \int d^3x \int_0^\beta d\tau\, e^{i\omega_n \tau} \int \langle \mathcal{T}_\tau(T_{21}(\boldsymbol{r},\tau)T_{21}(0))\rangle_0 \,, \tag{4.15}$$

where we applied a Wick rotation, $\tau = it$, and introduced the time-ordering symbol in imaginary time, $\mathcal{T}_\tau$. Note that while the underlying degrees of freedom are quarks, the relevant Matsubara frequencies are $\omega_n = 2\pi n$, since the fermions under the integral sign group together to form quantities with bosonic character: $\bar\psi(\cdot)\psi$. The global sign of $\Pi(\omega_n)$ is fixed by our sign convention for analytical continuations: $\Pi(\omega_n)|_{\omega_n \mapsto -i\omega(+\epsilon)} \mapsto -\Pi^{\mathrm{R}}(\omega)$, cf. Eq. (A.39) in the Appendix. The correlator $\Pi(\omega_n)$ is governed by non-perturbative physics resulting from the underlying interactions of the NJL model. We now apply a large-N_c expansion and organize this correlator in ring diagrams, ladder diagrams and higher-order terms:

$$\Pi(\omega_n) = \gamma_2 \;\begin{array}{c}\diagup\!\!\!\diagdown\end{array}\; \gamma_2 \;\; = \mathcal{O}(N_c^1) + \mathcal{O}(N_c^0) + \dots, \tag{4.16}$$

4. Microscopic theory of the shear viscosity

where the γ_2 matrices appear due to the evaluation of $T_{21} = i\bar{\psi}\gamma_2\partial_x\psi$, cf. Eqs. (4.14) and (4.15). Knowing the N_c-scaling of the NJL vertex $G \sim 1/N_c$, one finds at leading order $\mathcal{O}(N_c^1)$ a one-loop diagram contributing to the skeleton expansion of the four-point correlator $\Pi(\omega_n)$. The NJL Lagrangian in its simplest form in Eq. (3.40) takes only scalar and pseudoscalar but no vector or axialvector interactions into account: $\Gamma \in \{\mathbb{1}, i\gamma_5\}$. Iterating these interaction kernels in ring diagrams at leading order in $1/N_c$ to $\Pi(\omega_n)$ does not affect the correlator:

$$\gamma_2 \underbrace{\bigcirc \!\!\! \sim \!\!\! \bigcirc \!\!\! \sim \!\!\! \bigcirc}_{n \text{ loops}} \gamma_2 = 0, \qquad (4.17)$$

where $n = 1, 2, \dots$ in the diagram denotes the number of rings with interaction kernels Γ on both sides. Such ring diagrams are indeed vanishing, because the trace (in momentum and Dirac space) in the first ring vanishes due to the orthogonal operator structure involving the combination of γ_2 and Γ:

$$\begin{aligned}
T \sum_{n \in \mathbb{Z}} \int \frac{d^3p}{(2\pi)^3} \operatorname{Tr}\left[\gamma_2 G_\beta^F(\boldsymbol{p}, \nu_n) \Gamma G_\beta^F(\boldsymbol{p}, \nu_n)\right] = \\
= T \sum_{n \in \mathbb{Z}} \int \frac{d^3p}{(2\pi)^3} \frac{1}{(\nu_n^2 + \boldsymbol{p}^2 + m^2)^2} \operatorname{Tr}\left[\gamma_2 \Gamma m^2 + \gamma_2 \slashed{p}\Gamma\slashed{p} + \gamma_2 \slashed{p}\Gamma m + \gamma_2 \Gamma \slashed{p} m\right] = 0,
\end{aligned} \qquad (4.18)$$

where we have used the notation $\slashed{p} = \nu_n \gamma_4 - \boldsymbol{p} \cdot \boldsymbol{\gamma}$ and the full Matsubara propagator

$$G_\beta^F(\boldsymbol{p}, \nu_n) = \frac{\slashed{p} + m}{\nu_n^2 + \boldsymbol{p}^2 + m^2}, \qquad (4.19)$$

with frequencies $\nu_n = (2n+1)\pi T - i\mu$, cf. Appendix A.2. Exchange (ladder diagram) corrections to the chain in Eq. (4.17) are non-vanishing but of subleading order in $1/N_c$, because each rank in the ladder gives rise to a suppression factor $G^2 N_c \sim 1/N_c$. Note that adding one rank introduces two additional momentum integrations but only one additional color trace.

The shear viscosity in the NJL model has been derived previously in Refs. [FIO, FIO08a, FIO08b] using the Kubo formula, but setting the current-quark mass to zero: $m_0 = 0$. We point out that this result can in fact be derived without assuming to work in the chiral limit: to ensure the absence of iterated ring-diagram contributions it is indeed *not necessary* to assume this limit when taking only scalar and pseudoscalar interactions in the NJL Lagrangian into account. Iterated ring diagrams involving these interactions vanish naturally. Note that even in the chiral limit and the Nambu-Goldstone phase, the second term of the trace in Eq. (4.18), $\operatorname{Tr}\left[\gamma_2 \slashed{p}\Gamma\slashed{p}\right]$ would survive in the presence of vector interactions.

From the large-N_c analysis of the Kubo formula for the shear viscosity η we conclude that for all purely fermionic theories $\mathcal{L} = \mathcal{L}_{\text{kin}} + \mathcal{L}_{\text{int}}$ where all interactions are independent of the momentum and the coupling of the $2N$-vertex scales as $K_{2N} \sim N_c^{-(N-1)}$, the dominant contribution to the shear viscosity reads in Matsubara space:

$$\Pi(\omega_n) = \gamma_2 \bigcirc \gamma_2 + \mathcal{O}(N_c^0). \qquad (4.20)$$

We now proceed with the analytical evaluation of this leading-order contribution:

$$\Pi(\omega_n) = T \sum_{m \in \mathbb{Z}} \int \frac{d^3p}{(2\pi)^3} p_x^2 \operatorname{Tr}\left[\gamma_2 G_\beta^F(\boldsymbol{p}, \omega_n + \nu_m) \gamma_2 G_\beta^F(\boldsymbol{p}, \nu_m)\right] = \int \frac{d^3p}{(2\pi)^3} p_x^2 S(\boldsymbol{p}, \omega_n), \qquad (4.21)$$

4.1. Shear viscosity from Kubo formalism

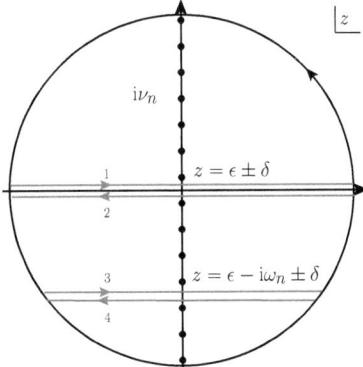

Figure 4.1.: Contour $\mathcal{C}$ and the four integration paths relevant for the evaluation of $S(\boldsymbol{p}, \omega_n)$. The poles located at the imaginary axes, $z = i\nu_n$ are the Matsubara poles coming from the Fermi distribution function $n_F^+(z)$, whereas the two branch cuts come from the fermion propagators describing thermal constituent quarks.

where we have denoted the integrand by $S(\boldsymbol{p}, \omega_n)$ as it is suggested in [FIO]. The trace refers to color, flavor, and Dirac space. We use residual calculus and interpret the sum $m \in \mathbb{Z}$ as sum over residua of the Fermi distribution function,

$$\operatorname{Res} n_F^+(z)\big|_{z=i\nu_n} = -T \,. \tag{4.22}$$

With the circle $\mathcal{C}$ in the complex plane centered at the origin with radius $R \to \infty$, we have therefore[32]:

$$S(\boldsymbol{p}, \omega_n) = -\frac{1}{2\pi i} \int_{\mathcal{C}} dz\, n_F^+(z) \operatorname{Tr}\left[\gamma_2 G_\beta^F(-iz + \omega_n) \gamma_2 G_\beta^F(-iz)\right] =$$
$$= -\frac{1}{2\pi i} \int_{-\infty}^{\infty} d\epsilon\, n_F^+(\epsilon) \operatorname{Tr}\left[\gamma_2 G_\beta^F(-i\epsilon + \omega_n) \gamma_2 G_\beta^F(-i\epsilon + \delta) - \gamma_2 G_\beta^F(-i\epsilon + \omega_n) \gamma_2 G_\beta^F(-i\epsilon - \delta) \right.$$
$$\left. + \gamma_2 G_\beta^F(-i\epsilon + \delta) \gamma_2 G_\beta^F(-\omega_n - i\epsilon) - \gamma_2 G_\beta^F(-i\epsilon - \delta) \gamma_2 G_\beta^F(-\omega_n - i\epsilon)\right]. \tag{4.23}$$

Apart from the Matsubara poles ν_n, the contour integral in the first line faces two branch cuts at $z = \epsilon - i\omega_n \neq i\nu_n$ and $z = \epsilon \neq i\nu_n$ as it can seen from the spectral representation

$$G_\beta(\boldsymbol{p}, \omega_n) = \int_{-\infty}^{\infty} d\epsilon\, \frac{\rho(\epsilon, \boldsymbol{p})}{\epsilon - i\omega_n} \,. \tag{4.24}$$

The contour integral separates into four contributions above and below the two branch cuts, denoted by the infinitesimal $\pm \delta \to 0$, respectively. The relative minus sign in Eq. (4.23) is due to the directions of ϵ-integration as it is sketched in Fig 4.1. Note that, actually, the second line therein has the prefactor $n_F^+(\epsilon - i\omega_n)$ but this equals due to the bosonic Matsubara frequencies $\omega_n = 2\pi nT$ just $n_F^+(\epsilon)$. Directly from its definition (4.24) we get

$$\lim_{\delta \to 0} \left[G_\beta(\boldsymbol{p}, -i\omega + \delta) - G_\beta(\boldsymbol{p}, -i\omega - \delta)\right] = 2\pi i\, \rho(\omega, \boldsymbol{p}) \,, \tag{4.25}$$

---
[32]In the following derivation we suppress the three-momentum in the notation writing $G_\beta^F(\cdot)$ instead of $G_\beta^F(\boldsymbol{p}, \cdot)$.

4. Microscopic theory of the shear viscosity

leading to

$$S(\boldsymbol{p}, \omega_n) = -\int_{-\infty}^{\infty} d\epsilon\, n_F^+(\epsilon)\, \text{Tr}\left[\gamma_2 \rho(\epsilon)\gamma_2 \left(G_\beta^F(\omega_n - i\epsilon) + G_\beta^F(-\omega_n - i\epsilon)\right)\right]. \quad (4.26)$$

For the shear viscosity η we need the imaginary part of the retarded correlator $\Pi^R(\omega)$, therefore we perform at this stage for $S(\boldsymbol{p}, \omega_n)$ the analytical continuation to Minkowski space, $\omega_n \mapsto -i\omega + \delta$, and find when taking its imaginary part and sending $\delta \to 0$:

$$\text{Im}\, S(\boldsymbol{p}, -i\omega) = -\int_{-\infty}^{\infty} d\epsilon\, n_F^+(\epsilon)\, \text{Tr}\left[\gamma_2 \rho(\epsilon)\gamma_2\, \text{Im}\left(G_\beta^F(-i(\omega + \epsilon) + \delta) + G_\beta^F(i(\omega - \epsilon) - \delta)\right)\right] =$$

$$= -\pi \int_{-\infty}^{\infty} d\epsilon\, n_F^+(\epsilon)\, \text{Tr}\left[\gamma_2 \rho(\epsilon)\gamma_2 \left(\rho(\epsilon + \omega) - \rho(\epsilon - \omega)\right)\right], \quad (4.27)$$

where we have used the relation (cf. Eq. (3.108))

$$\lim_{\delta \to 0} \text{Im}\, G_\beta(\boldsymbol{p}, -i\omega \pm \delta) = -\lim_{\delta \to 0} \text{Im}\, G_{R/A}(\omega, \boldsymbol{p}) = \pm\pi\rho(\omega, \boldsymbol{p}). \quad (4.28)$$

Shifting for the second term in the integrand the ϵ-integration my $\epsilon \mapsto \epsilon + \omega$ we find as final answer in Minkowski space:

$$\text{Im}\, S(\boldsymbol{p}, -i\omega) = -\pi \int_{-\infty}^{\infty} d\epsilon \left(n_F^+(\epsilon) - n_F^+(\epsilon + \omega)\right) \text{Tr}\left[\gamma_2 \rho(\epsilon)\gamma_2 \rho(\epsilon + \omega)\right]. \quad (4.29)$$

In order to calculate finally the shear viscosity we combine Eqs. (4.13) and (4.21) and get

$$\eta = -\int \frac{d^3 p}{(2\pi)^3} p_x^2 \left.\frac{d}{d\omega} \text{Im}\, S(\boldsymbol{p}, -i\omega)\right|_{\omega=0} =$$

$$= \frac{\pi}{T} \int_{-\infty}^{\infty} d\epsilon \int \frac{d^3 p}{(2\pi)^3} p_x^2\, n_F^+(\epsilon)\left(1 - n_F^+(\epsilon)\right) \text{Tr}\left[\gamma_2\, \rho(\epsilon, \boldsymbol{p})\, \gamma_2\, \rho(\epsilon, \boldsymbol{p})\right], \quad (4.30)$$

where the combination of Fermi distribution functions arises from their derivatives:

$$\frac{dn_F^\pm(\epsilon)}{d\epsilon} = \beta n_F^\pm(\epsilon)(n_F^\pm(\epsilon) - 1) < 0. \quad (4.31)$$

This is the main result for the shear viscosity in the NJL model from Kubo formalism. In the subsequent Section 4.2 we will study the shear viscosity assuming a quark self-energy that can be described by a parameterized spectral width Γ. Later, in Section 6.2, we will extend this discussion and evaluate the master formula (4.30) with full results for the quark self-energy calculated from the NJL model.

We would like to emphasize two properties of the shear viscosity η. First, though the integrand of η contains only matter contributions from quark distribution function n_F^+, the shear viscosity is invariant under $\mu \mapsto -\mu$. This follows from the fact that the ϵ-integration ranges over both positive and negative energies. One also uses $n_F^+(-\epsilon)\left(1 - n_F^+(-\epsilon)\right) = n_F^-(\epsilon)\left(1 - n_F^-(\epsilon)\right)$ and that fact that $\text{Tr}\left[\gamma_2 \rho(\epsilon, \boldsymbol{p})\gamma_2 \rho(\epsilon, \boldsymbol{p})\right]$ is an even function in ϵ, cf. Appendix A.3. The second comment refers to the off-shell structure of the Kubo formula. The ϵ and p-integrations are carried out independently, therefore one has to provide the full spectral function without using on-shell restrictions. In Section 6.2 we will discuss in detail the differences between the on- and off-shell treatments of the master formula (4.30). For the proceeding parameter study, the spectral width $\Gamma(p)$ is a function of momentum $p = |\boldsymbol{p}|$ only, i.e. there we restrict to the on-shell approximation of the Kubo formula only.

4.2. Parameter study of the shear viscosity

In this section we discuss the shear viscosity $\eta[\Gamma]$ of the NJL model at leading-order in a large-N_c expansion systematically. Working first with the unphysical assumption of a momentum-independent spectral width we are able to derive analytical results and some convergence criterion for $\eta[\Gamma]$. Based on this rough assumption we introduce a numerical scheme which allows us to treat more efficiently also the physical cases where the spectral width depends on the quark momentum. Not surprisingly, we find a very strong cutoff dependence of the final results for η that, however, ensures physical results. The non-perturbative structure of the NJL model implies that ladder-diagram resummation does not affect the resulting shear viscosity drastically. Finally, since we are actually interested in the thermal dependence of the shear viscosity, effects induced by the thermal quark mass are investigated.

We work in a quasi-particle approximation and assume that the quark self-energy can be expressed by a single parameter, the spectral width $\Gamma(p)$. We use the following ansatz [FIO, FIO08a]:

$$G_{\text{R/A}}(p_0, \boldsymbol{p}) = \frac{1}{\not{p} - m \pm i\,\text{sgn}(p_0)\Gamma(p)}\,, \tag{4.32}$$

where we formally substituted the pole description $\epsilon \mapsto \Gamma$ in Eq. (A.32). Note that for positive energies, $p_0 > 0$, the retarded propagator G_R is just the Feynman propagator. The thermal environment affects parametrically not only the spectral width $\Gamma(p;T,\mu)$, but also, via the gap equation (3.83), the quark mass $m(T,\mu)$. From this ansatz the spectral function is derived using the "inverse relation" Eq. (3.108):

$$\rho(p_0, \boldsymbol{p}) = -\frac{1}{\pi}\text{Im}\,G_R(p_0, \boldsymbol{p}) = \frac{\text{sgn}(p_0)\Gamma(p)}{\pi}\frac{p^2 + 2m\not{p} + m^2 + \Gamma^2}{(p^2 + m^2 + \Gamma(p)^2)^2 - 4m^2 p^2}\,, \tag{4.33}$$

where $p^2 = p_0^2 - \boldsymbol{p}^2$ and the denominator is denoted for convenience by

$$X(p) = \left(p^2 + m^2 + \Gamma(p)^2\right)^2 - 4m^2 p^2 = \left(p^2 - m^2 + \Gamma(p)^2\right)^2 + 4m^2 \Gamma(p)^2\,. \tag{4.34}$$

For the shear viscosity (4.30) one easily carries out the color and flavor traces in Eq. (4.30), resulting in trivial prefactors N_c and N_f, respectively. The trace in Dirac space gives

$$\text{Tr}[\gamma_2 \rho \gamma_2 \rho] = \frac{\Gamma(p)^2}{\pi^2 X(p)^2}\,\text{Tr}\left[\gamma_2(p^2 + 2m\not{p} + m^2 + \Gamma(p)^2)\gamma_2(p^2 + 2m\not{p} + m^2 + \Gamma(p)^2)\right] =$$

$$= \frac{\Gamma(p)^2}{\pi^2 X(p)^2}\,\text{Tr}\left[\gamma_2^2(X(p) + 4p^2 m^2) + 4m^2 \gamma_2 \not{p} \gamma_2 \not{p}\right] = \tag{4.35}$$

$$= \frac{4\Gamma(p)^2}{\pi^2 X(p)^2}\left[8m^2 p_y^2 - X(p)\right] \longrightarrow \frac{32 m^2 p_y^2 \Gamma^2(p))}{\pi^2 X(p)^2} \quad \text{as } X \to 0\,,$$

where we have used $\text{Tr}[\gamma_2 \not{p}\gamma_2\not{p}] = 4p^2 + 8p_y^2$. The discussion in Section 4.2.1 will show that the shear viscosity is dominated by regions where the denominator $X(p)$ becomes small. Therefore we have taken for $\text{Tr}[\gamma_2 \rho \gamma_2 \rho]$ only the dominant term $\sim X(p)^{-2}$ into account and dropped the term $\sim X(p)^{-1}$. For an isotropic medium the spectral width $\Gamma(p)$ does not depend on any angle. The angular part of the $\text{d}^3 p$ integration can be readily carried out using

$$\int \text{d}\Omega\, p_i^2 p_j^2 = \frac{4p^4 \pi}{15}\,, \quad \text{for } i \neq j\,, \tag{4.36}$$

where again the factor 15 appears as we have discussed already at the beginning of this section

73

4. Microscopic theory of the shear viscosity

in Eq. (4.2). The shear viscosity becomes:

$$\eta[\Gamma(p)] = \frac{16\beta N_c N_f}{15\pi^3} \int_{-\infty}^{\infty} d\epsilon \int_0^{\infty} dp\, p^6 \frac{m^2\, \Gamma(p)^2\, n_F^+(\epsilon)(1 - n_F^+(\epsilon))}{[(\epsilon^2 - p^2 - m^2 + \Gamma(p)^2)^2 + 4m^2\Gamma(p)^2]^2}\,. \tag{4.37}$$

This is a main result for the shear viscosity from Kubo formalism assuming the parameterization of the dressed quark propagator as given in Eq. (4.32). The advantage of this parameterization is that only a single (momentum dependent) spectral width, $\Gamma(p)$, describes the shear viscosity. In the following we study the result (4.37) in detail in order to explore the qualitative and quantitative behavior of the shear viscosity, e.g. for different parameterizations of $\Gamma(p)$.

4.2.1. Analytical results for a constant spectral width

To start with we assume a spectral width $\Gamma = $ const., independent of the momentum and any thermal parameters. This allows for an analytical result for the momentum integral in Eq. (4.37), therefore one is left with the one-dimensional numerical ϵ-integration only. We use

$$\int_0^{\infty} dp\, \frac{p^6}{[(A - p^2)^2 + B^2]^2} = \frac{\pi}{8\sqrt{2}} \frac{\sqrt{\sqrt{A^2 + B^2} - A}}{B^4} \left[(2A^2 + 3B^2)\sqrt{A^2 + B^2} + 2A(A^2 + 2B^2)\right], \tag{4.38}$$

where we identifies $A = \epsilon^2 - m^2 + \Gamma^2$ and $B = 2m\Gamma$. One can derive this result by extending the integration range to $p \in \mathbb{R}$ (the integrand is an even function of p) and using residual calculus. Having performed the p-integration analytically by hand improves the computation time for one value of η by roughly one order of magnitude. Furthermore, it helps finding an appropriate approximation scheme for the whole (ϵ, p)-integration when the spectral width is momentum dependent.

In Fig. 4.2 we show the results for $\eta[\Gamma; T, \mu]$ assuming a constant spectral width and using a constant quark mass of $m = 100$ for convenience. For $\Gamma \to 0$ the shear viscosity diverges, as it is expected from general considerations. This limit describes a system of free quarks for which the mean free path is infinite. With increasing temperature and quark chemical potential, the shear viscosity increases, but the dependence on temperature is more pronounced. Compare these figures to those in Ref. [FIO], where $\eta[\Gamma]$ has been evaluated numerically without a momentum-space cutoff, equivalent to our analytical approach based on Eq. (4.38), i.e. $p \in (0, \infty)$ which is actually unphysical in the NJL model, since the model becomes non-interacting for too high momenta leading to $\Gamma \to 0$ implying a divergent shear viscosity. However, inspecting the detailed behavior of the integrand in Eq. (4.37), a convergence criterion for the shear viscosity in the absence of a momentum-space cutoff can be derived:

In order for the shear viscosity $\eta[\Gamma]$ as functional of $\Gamma(p)$ to be convergent, the asymptotic $\Gamma(p)$ should not converge too rapidly to zero:

$$\eta[\Gamma(p)] < \infty \quad \Leftrightarrow \quad p^{7/2} e^{-\beta p/2} \in o(\Gamma(p))\,, \tag{4.39}$$

where o($\cdot$) denotes the little Landau symbol[33]. Possible parameterizations of $\Gamma(p)$ satisfying this

---

[33]The notation $f \in o(g)$ is used to express accurately that "f is growing less fast than g", meaning that $f(x)/g(x) \to 0$ for $x \to \infty$. More intuitively, this also means that "g grows much faster than f".

4.2. Parameter study of the shear viscosity

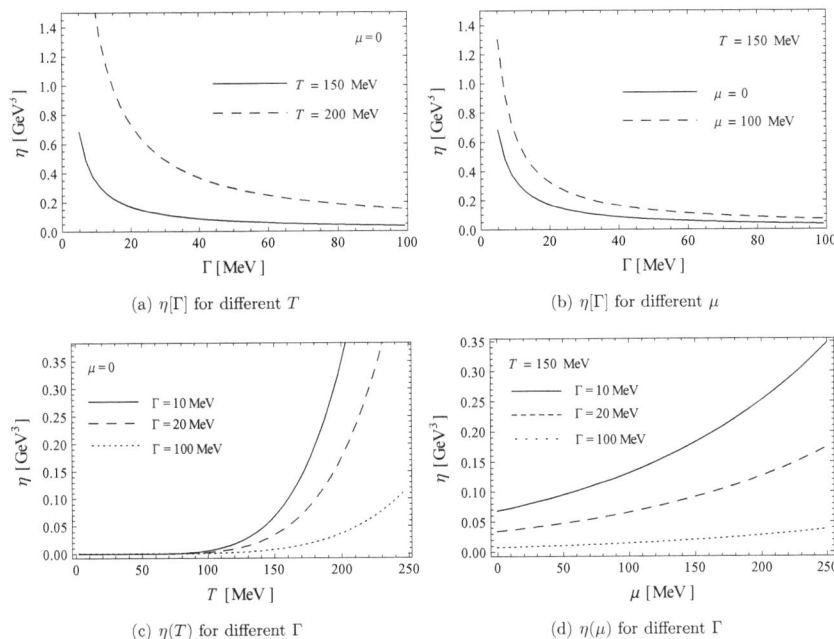

Figure 4.2.: Shear viscosity η as function of a constant spectral width Γ, temperature T and quark chemical potential μ. In this analytical calculation no momentum cutoff was used.

constraint are:

$$\begin{aligned}
\text{constant}: \quad & \Gamma_{\text{const}} = 100 \text{ MeV}, \\
\text{exponential}: \quad & \Gamma_{\exp}(p) = \Gamma_{\text{const}}\, e^{-\beta p/8}, \\
\text{Lorentzian}: \quad & \Gamma_{\text{Lor}}(p) = \Gamma_{\text{const}}\, \frac{\beta p}{1+(\beta p)^2}, \\
\text{divergent}: \quad & \Gamma_{\text{div}}(p) = \Gamma_{\text{const}}\, \sqrt{\beta p}.
\end{aligned} \qquad (4.40)$$

Note that all these parameterizations lead to a finite shear viscosity and no mathematical regularization must be applied. A Gaussian shape, $\Gamma \sim \exp\left(-\beta^2 p^2\right)$, for instance does not satisfy the convergence criterion (4.39). The particular shapes of the prototype widths (4.40) have been chosen because of their different behavior at small and large momenta: vanishing or non-vanishing $\Gamma(p=0)$, convergent or divergent $\Gamma(p)$ for $p \to \infty$. These prototypes represent physical spectral widths in several theories [Lan10, LKW12]: $\Gamma(p)$ in ϕ^4 theory, for instance, is a monotonic function and converges to zero for large momenta. This can be described by the Lorentz parameterization for large momenta: $\lim_{p \to \infty} \Gamma_{\text{Lor}}(p) \sim T/p$. In contrast, the spectral width of an interacting pion gas diverges for $p \to \infty$.

4.2.2. Numerical approach to momentum-dependent spectral widths

The numerical treatment of $\eta[\Gamma(p)]$ in Eq. (4.37) with a momentum-dependent spectral width is based on the observation that its integrand ranges typically over something like ten orders of

4. Microscopic theory of the shear viscosity

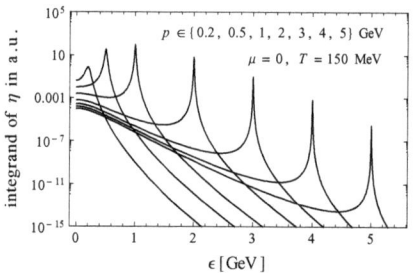

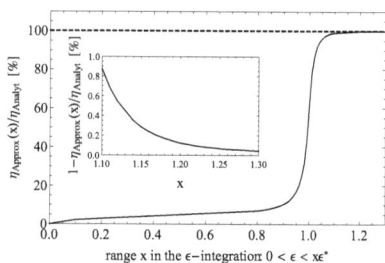

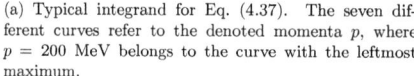

(a) Typical integrand for Eq. (4.37). The seven different curves refer to the denoted momenta p, where $p = 200$ MeV belongs to the curve with the leftmost maximum.

(b) Convergence plot of the numerical approximation scheme for $\eta[\Gamma]$ shown for a constant spectral width compared to its analytical result that is reproduced exactly only for $x \to \infty$.

Figure 4.3.: Details on the numerical approximation scheme: due to the sharp peak structure of the integrand as shown in panel (a), the ϵ-integration in Eq. (4.37) is restricted to $|\epsilon| < x\epsilon^*$, where $x = 1.3$ provides an accuracy of 10^{-4} as it is shown in panel (b). See also the discussion in the text.

magnitude as it is shown in Fig 4.3(a). We have extended the values of p and ϵ to unphysically high values in order to clearly display the underlying systematics. For every momentum p there is a maximum of the integrand, located at

$$\epsilon^*(p) = \sqrt{p^2 + m(T,\mu)^2 - \Gamma(p;T,\mu)^2} \longrightarrow p \quad \text{for large momenta } p \, . \qquad (4.41)$$

For all physical momenta, i.e. empirically for $p \lesssim 2-3$ GeV which is a bound far above the NJL-cutoff scale ($\Lambda < 1$ GeV), the integrand is most sizable within the vicinity of $\epsilon^*(p)$. Adaptive methods for numerical integration run into serious trouble when facing a sharp peak structure as present for the Kubo formula for $\eta[\Gamma(p)]$: either the step size becomes too tiny for fast convergence (or convergence at all), or the most important contribution in the vicinity of the peak is not sampled by a too coarse step size. We overcome this numerical issue by cutting the ϵ-integration "by hand" and allow only for $|\epsilon(p)| < x\epsilon^*(p)$ for some $x \gtrsim 1$. As it is found from Fig. 4.3(b) the choice $x = 1.3$ is sufficient to reproduce the analytical result for a constant spectral width up to a relative error of 10^{-4}. It can be seen how in the vicinity of the maximizer $\epsilon \approx \epsilon^*(p)$, i.e. $x \approx 1$, the dominant contributions to the shear viscosity are integrated. We report that for all momentum-dependent parameterizations $\Gamma(p)$ defined in Eq. (4.40) the integrands look qualitatively the same as for a constant spectral width which has been used in Fig. 4.3. Therefore we expect the described numerical scheme to work well also in these and more physical cases as we will calculate in Chapter 5.

4.2.3. Cutoff dependence

Generally, the shear viscosity increases when the spectral width decreases, cf. Fig 4.2(a). This behavior is also visible in Fig. 4.4(a) when comparing the prototype parameterizations of $\Gamma(p)$: the "more divergent" the spectral width as $p \to \infty$, the smaller the corresponding shear viscosity:

$$\eta_{\text{Lor}} > \eta_{\text{exp}} > \eta_{\text{const}} > \eta_{\text{div}} \, , \qquad (4.42)$$

4.2. Parameter study of the shear viscosity

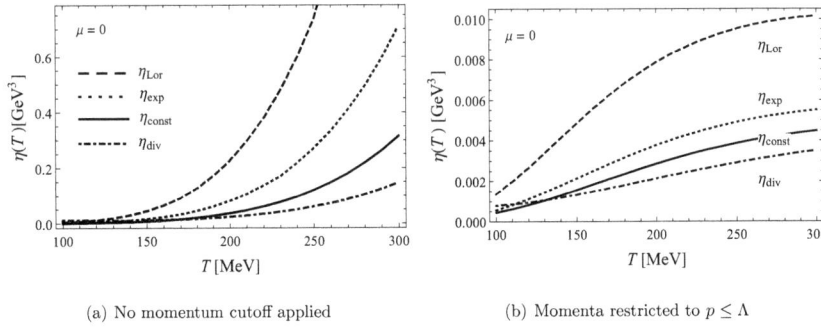

(a) No momentum cutoff applied

(b) Momenta restricted to $p \leq \Lambda$

Figure 4.4.: Shear viscosity η as function of temperature at vanishing quark chemical potential, for different schematic parameterizations (4.40) of the spectral width $\Gamma(p)$. Sequence of curves and qualitative change from scenarios without (a) and with (b) momentum cutoff $\Lambda = 651\,\text{MeV}$ are discussed in the text.

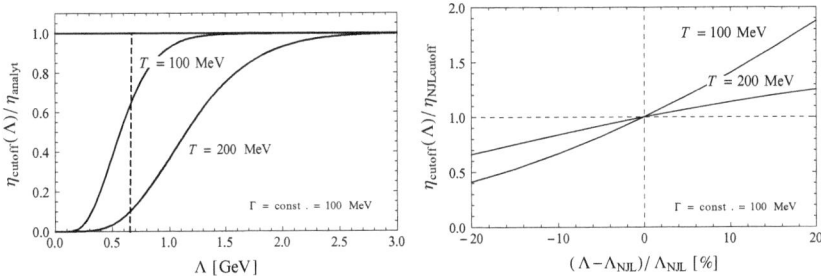

(a) Absolute cutoff dependence; the vertical dashed line represents the position of the physical NJL cutoff

(b) Relative cutoff dependence around $\Lambda = 651\,\text{MeV}$

Figure 4.5.: Absolute (a) and relative (b) cutoff dependence of the shear viscosity, demonstrating the suppression of high-momentum contributions when the standard (physical) NJL cutoff is used. The plots are drawn at $\mu = 0$ and for constant spectral width $\Gamma = 100\,\text{MeV}$.

using notations as in Eq. (4.40). This sequence is implied by the corresponding (inverse) order for the spectral widths. These arguments hold also for non-vanishing quark chemical potentials. Assuming the spectral width itself to be independent of the chemical potential, the shear viscosity increases for increasing μ, but the qualitative shape of $\eta(T)$ does not change. We note that the results in Fig. 4.4 have been derived using a constant constituent quark mass, $m = 325\,\text{MeV}$ being its "physical" vacuum value.

The integrand of $\eta[\Gamma(p)]$, Eq. (4.37), is sizable for unphysically large momenta, so we expect a strong cutoff dependence. In the NJL model the quasiparticle interactions are restricted to quark momenta $p \leq \Lambda = 651\,\text{MeV}$. Quarks with momenta $p > \Lambda$ do not interact and have infinite mean free paths. Restricting the momentum integration to the interval $p \leq \Lambda$, we find a shear viscosity as shown in Fig. 4.4(b). Excluding $p > \Lambda$ reduces the shear viscosity by one

77

4. Microscopic theory of the shear viscosity

order of magnitude at low temperatures and even by two orders of magnitude at high T. As expected, this expresses a very strong cutoff dependence. In addition to these quantitative differences, the qualitative behavior of the shear viscosity also changes strongly and flattens for high temperatures.

This strong cutoff dependence is investigated in more detail in Fig. 4.5: the contributions taken into account (compared to the analytical result for η) depend strongly on temperature and just weakly on the quark chemical potential. At $T = 200\,\text{MeV}$ the momentum cutoff excludes about 90% of the full integral extended to infinity, see Fig. 4.5(a). As shown in Fig. 4.5(b), varying the cutoff by up to $\pm 20\%$ implies for η a change of up to 100%.

To assess the order of magnitude of the NJL shear viscosity, a comparison with $\eta(T)$ for other systems is instructive. For example, an interacting pion gas treated within the framework of chiral perturbation theory [LKW12] has a typical shear viscosity of order $\eta(T) \approx 40\,\text{MeV}/\text{fm}^2 \approx 1.6 \cdot 10^{-3}\,\text{GeV}^3$ at $T \approx 100\,\text{MeV}$. This is a similar order of magnitude as the results shown in Fig. 4.4(b) when applying the NJL cutoff $\Lambda = 651\,\text{MeV}$. We recall that this cutoff is fixed by reproducing physical observables such as the pion decay constant in vacuum and not adjusting the overall scale of shear viscosity. In contrast, a physically meaningful order of magnitude for η follows naturally when incorporating the NJL cutoff.

4.2.4. Perturbative aspects and ladder-diagram resummation

As we have already mentioned the shear viscosity η diverges for non-interacting systems, i.e. for a vanishing spectral width, corresponding to infinite mean free path. Close to this limit η can be expanded in a Laurent series (as realized for example analytically in ChPT and $\lambda \phi^4$ theory [LKW12]):

$$\eta[\Gamma] = \frac{A_{-1}}{\Gamma} + A_0 + A_1 \Gamma + A_2 \Gamma^2 + \ldots \quad (4.43)$$

For small Γ, the combination $\eta \cdot \Gamma$ is just the residue A_{-1}. What does "small" mean in this context? In contrast to the perturbative $\lambda \phi^4$ theory where $\Gamma \sim \lambda^2$, the NJL model is generically non-perturbative in its coupling, even though the scaling $G \sim 1/N_c$ applies. The spectral width is therefore not expected to be sufficiently small in order to permit an expansion as in Eq. (4.43). Fig. 4.6 shows results of the fully non-perturbative calculation of $\eta \cdot \Gamma$ as a function of the inverse width, conveniently written as $x = m_\pi/\Gamma$, at different T and μ in comparison with the residue

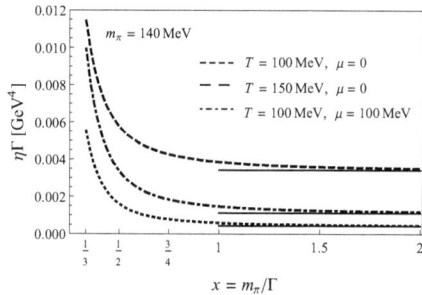

Figure 4.6.: Scaling of $\eta \cdot \Gamma$ for different T and μ as function of the inverse width expressed in units of the pion mass m_π. Solid horizontal lines correspond to the residues A_{-1} of $\eta[\Gamma]$ in Eq. (4.43). A constituent quark mass $m = 100\,\text{MeV}$ has been used for convenience.

A_{-1}. As it can be seen from the figure, corrections to the leading term of the Laurent series (4.43) are small for $x > 1.5$ (demanding 10% accuracy or better). From these considerations we conclude that a perturbative approach is justified only for spectral widths $\Gamma \ll , m_\pi = 140$ MeV.

The discussion of a perturbative treatment of $\eta[\Gamma(p)]$ is closely related to the resummation of ladder diagrams: if in the large-N_c limit the spectral width decreases, i.e. $\Gamma \sim 1/N_c$ as suggested by hot-QCD calculations [Def05] where the coupling $\alpha_s \sim 1/N_c$ becomes small, then the perturbative regime is reached in this limit and the Laurent series expansion in (4.43) can be restricted to its leading-order term. As seen from Eq. (4.37), for a constant but small spectral width $\Gamma \to 0$ the residue A_{-1} can be identified with the remaining ϵ-integral:

$$\eta[\Gamma(p)] \longrightarrow \frac{2N_c N_f}{15\pi^2 T} \int_{|\epsilon|>m} d\epsilon \, \frac{(\epsilon^2 - m^2)^{5/2} \, n_F^+(\epsilon)(1 - n_F^+(\epsilon))}{m\, \Gamma(\sqrt{\epsilon^2 - m^2})} \,. \qquad (4.44)$$

In contrast to the non-perturbative result in Eq. (4.37) the ϵ-integration excludes the region $|\epsilon| < m$. This is due to the delta functions appearing in the limit of small Γ. The momentum integration of the integrand involving $\delta(\epsilon^2 - p^2 - m^2)$ is readily carried out. Compare this expression with the results from a perturbative treatment in [HK11]. Because in the limit $\Gamma \to 0$ only the residue term of $\eta[\Gamma(p)]$ is relevant, ladder diagrams now become sizable corrections and contribute also at leading order. Furthermore, the shear viscosity now scales as $\eta \sim N_c^2$ and no longer linearly with N_c as Eqs. (4.30) and (4.37) do for N_c-independent spectral function ρ and width Γ, respectively.

We conclude that ladder diagram resummation is necessary in the perturbative regime of $\eta[\Gamma(p)]$ in Eq. (4.37), i.e. when the spectral width is small, $\Gamma \ll m_\pi$. In the NJL model with its genuine non-perturbative structure, the physical spectral width is large and outside the perturbative regime. This will be demonstrated by an explicit calculation in Chapter 5. Therefore, contributions from ladder diagram resummation are subleading corrections, while the shear viscosity functional (4.37) is valid also for large spectral width when including all orders of the Laurent series expansion (4.43).

4.2.5. Effects of thermal quark masses on the shear viscosity

In this parameter study the constituent-quark mass has so far been treated as a constant. We now proceed to incorporate its explicit T and μ dependence as we have calculated it from the Hartree-Fock gap equation (3.75) with the results shown in Fig. 3.4. In Fig. 4.7(a) we show the shear viscosity η for varying constituent-quark mass m where the mass is treated as a parameter and assuming a constant spectral width $\Gamma_{\text{const}} = 100$ MeV. For $m \to 0$ the shear viscosity becomes divergent, again due to *pinched poles* appearing in Eq. (4.37) in this limit. In fact, the origin of this divergence is the same as for $\Gamma \to 0$, since m and Γ are formally (almost) interchangeable in the integrand of Eq. (4.37). For large constituent quark masses, two effects occur: first, the maximizer $\epsilon^*(p) \sim m$ (4.41) moves to larger values, and second, the integrand of Eq. (4.37) scales as m^{-6}. Both features result in a decreasing function $\eta(m)$.

Taking the full thermal dependence of the constituent-quark mass into account has an essential influence on the shear viscosity, see Fig. 4.7(b): for small T, a constant mass $m = 325$ MeV approximates the thermal constituent quark mass. In contrast, at large T, with a melting chiral condensate, the dropping dynamical quark mass implies a strongly increasing shear viscosity, qualitatively different from the case with constant quark mass.[34] From this study one can conclude that besides the NJL cutoff the thermally generated constituent-quark mass dominates both the qualitative and quantitative result for the shear viscosity $\eta(T)$.

---

[34]We have chosen to compare thermal and non-thermal results for constant and exponential parameterizations of the spectral width. For Γ_{Lor} and Γ_{div} the results are qualitatively similar.

4. Microscopic theory of the shear viscosity

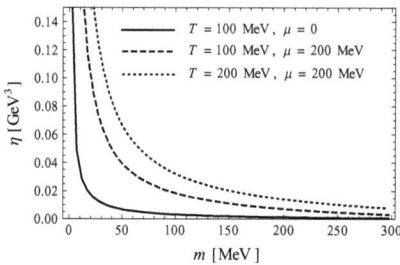
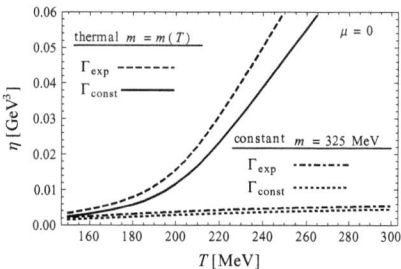

(a) Parametric dependence on the dynamical constituent quark mass m for a constant spectral width $\Gamma = 100$ MeV

(b) Shear viscosity with thermal quark mass $m(T)$ at vanishing quark chemical potential compared to a constant (vacuum) quark mass

Figure 4.7.: Dependence of the shear viscosity on the constituent-quark mass: (a) parametric dependence on m assuming a constant spectral width $\Gamma = 100$ MeV; (b) temperature dependence of the shear viscosity including the full thermal constituent-quark masses $m(T, \mu)$ (upper lines, i.e. without dots) compared to a constant quark mass $m = 325$ MeV (lower lines, i.e. with dots). The results can be understood from Fig. 3.4.

4.3. Kinetic theory

So far we have intensely discussed and applied the Kubo formalism for evaluating the shear viscosity. There is an alternative approach, the kinetic theory using the Boltzmann equation, which is widely used when investigating transport coefficients. It can be used for the description of the dynamics of a fluid composed of quasiparticles which is not too far from equilibrium. In this section we illustrate the derivation of the general expression for the shear viscosity within this formalism and compare it to the results from Kubo formalism derived in Section 2.3. We start with the Boltzmann equation, cf. for instance [HK85, CEM13]:

$$\frac{\partial f}{\partial t} = \underbrace{-\boldsymbol{v} \cdot \frac{\partial f}{\partial \boldsymbol{x}}}_{\text{diffusion}} \underbrace{-\boldsymbol{F} \cdot \frac{\partial f}{\partial \boldsymbol{p}}}_{\text{external}} + \left.\frac{\partial f}{\partial t}\right|_{\text{collisions}} . \tag{4.45}$$

Its principal shape does not differ for fermionic and bosonic systems, so $f(\boldsymbol{x}, \boldsymbol{p})$ denotes one of the corresponding distribution functions. It describes the phase space (probability) density of the quasiparticles that the fluid consists of. At equilibrium one has the distribution functions

$$f_0^\pm(\boldsymbol{x}, \boldsymbol{p}) = n_{\text{F/B}}^\pm(u_\mu p^\mu) = \frac{1}{e^{\beta(E - \boldsymbol{p} \cdot \boldsymbol{u} \mp \mu)} \pm 1} , \tag{4.46}$$

with the quasiparticle four momentum $p^\mu = (E, \boldsymbol{p})$ and the four velocity of the fluid $u^\mu(\boldsymbol{x})$ as it has been already defined in Eq. (2.62). The first term in the Boltzmann equation (4.45) describes diffusion processes, i.e. the spatial variation of the non-equilibrium distribution function f, whereas the second term describes external forces, e.g. a gravitational field or electromagnetism effects. The collision term is usually[35] treated in the so-called *relaxation-time approximation*,

---

[35] We mention that there are several approximation schemes on can apply to the Boltzmann equation. Apart from the relaxation-time approximation which uses an ansatz for the collision term, one could also apply variational methods like the Chapman-Enskog approximation, cf. [PPSG12, WP12] and the references therein.

4.3. Kinetic theory

i.e. one makes the ansatz [SR09]:

$$\left.\frac{\partial f}{\partial t}\right|_{\text{collisions}} = -\frac{\delta f}{\tau} . \qquad (4.47)$$

One thereby assumes that the system's deviation from equilibrium, $\delta f = f - f_0$, is small compared to the relaxation time τ, which is equivalent to $\langle A \rangle \gg \langle B \rangle$ in linear response theory, cf. Eq. (2.69). We recall that the operator $A = \beta H$ denotes the equilibrium part of the thermodynamic system (with Hamiltonian H) and the operator B describes deviations from it.

Usually, τ is determined by calculating cross sections from totally elastic $2 \to 2$ scattering processes as it is done for instance in [SR10]. The relativistic generalization of Eq. (4.47) reads at leading order in δf:

$$(p \cdot \partial) f_0 = -\frac{E}{\tau} \delta f . \qquad (4.48)$$

Defining the energy-momentum tensor directly from the Lagrangian as it was done before in Eq. (4.14) is not useful for the hydrodynamic description. Instead, one defines its thermal average (with respect to the equilibrium state) as a three-momentum integral over the distribution functions[36]:

$$T^{\mu\nu} = \int \frac{\mathrm{d}^3 p}{(2\pi)^3} \frac{p^\mu p^\nu}{E} \left(f^+ + f^- \right) . \qquad (4.49)$$

From this, using the relaxation-time ansatz, the linear perturbation of the energy-momentum tensor in δf can be derived as:

$$\tau^{\mu\nu} = \delta T^{\mu\nu} = -\int \frac{\mathrm{d}^3 p}{(2\pi)^3} \frac{p^\mu p^\nu}{E^2} (p \cdot \partial) \left(\tau f_0^+ + \bar{\tau} f_0^- \right) , \qquad (4.50)$$

with $\bar{\tau}$ denoting the mean life time of the anti-quasiparticle. We identify this linear correction to the energy-momentum tensor with the dissipative tensor $\tau^{\mu\nu}$ introduced in the Kubo formalism in Eq. (2.63). Again, only first-order derivatives in the dissipative force have been taken into account.

In the following we derive expressions for the time evolution of the thermal parameters T and μ. For this we need the following two Maxwell relations,

$$\frac{\partial V}{\partial N} = \frac{\partial \mu}{\partial P} , \quad \text{and} \quad \frac{\partial S}{\partial V} = \frac{\partial P}{\partial T} , \qquad (4.51)$$

where the first one can be derived from the enthalpy differential

$$\mathrm{d}H = T \,\mathrm{d}S + V \,\mathrm{d}P + \mu \,\mathrm{d}N , \qquad (4.52)$$

and the second relation from the free-energy differential

$$\mathrm{d}F = -S \,\mathrm{d}T - P \,\mathrm{d}V + \mu \,\mathrm{d}N . \qquad (4.53)$$

Starting from the internal energy $E(S, V, N)$ we derive using Eqs. (4.51) the thermodynamic relation for the energy density:

$$\mathrm{d}E = T \,\mathrm{d}S - P \,\mathrm{d}V + \mu \,\mathrm{d}N$$
$$\Leftrightarrow \quad \epsilon = \frac{\mathrm{d}E}{\mathrm{d}V} = T \frac{\partial S}{\partial V} - P + \mu \frac{\partial N}{\partial V} = T \frac{\partial P}{\partial T} - P + \mu \frac{\partial P}{\partial \mu} . \qquad (4.54)$$

---

[36] The integrand's shape of $T^{\mu\nu}$ can be considered as combination of a Lorentz-covariant integral measure, $\mathrm{d}^3 p / E$, the most general second-rank tensor made out of four-momenta, $p^\mu p^\nu$, followed by the phase space density functions, $f^\pm$.

4. Microscopic theory of the shear viscosity

From this we see that the energy density can be considered as a function of temperature and quark chemical potential only, $\epsilon(T,\mu)$, since the pressure is $P(T,\mu)$. Using again the first Maxwell relation we find the analogous statement for the particle density,

$$n = \frac{\partial N}{\partial V} = \frac{\partial P}{\partial \mu}, \qquad (4.55)$$

which implies $n(T,\mu)$. Inverting both functions, ϵ and n, temperature and chemical potential are determined when knowing the energy and particle density: $T(\epsilon,n)$ and $\mu(\epsilon,n)$. Therefore the total time derivative of the pressure can be expressed in two equivalent ways:

$$\frac{\mathrm{d}P(T,\mu)}{\mathrm{d}t} = \frac{\partial P}{\partial T}\frac{\partial T}{\partial t} + \frac{\partial P}{\partial \mu}\frac{\partial \mu}{\partial t} \stackrel{!}{=} \frac{\partial P}{\partial \epsilon}\frac{\partial \epsilon}{\partial t} + \frac{\partial P}{\partial n}\frac{\partial n}{\partial t} = \frac{\mathrm{d}P(\epsilon,n)}{\mathrm{d}t}. \qquad (4.56)$$

Having an explicit expression for the right-hand side of this identity, the time evolution of T and μ can be extracted. This can be achieved using energy density conservation:

$$0 = u_\nu \partial_\mu T^{\mu\nu} =$$
$$= u_\nu \left(\partial_\mu \epsilon + \partial_\mu P\right) u^\mu u^\nu + u_\nu \left(\epsilon + P\right)\left[(\partial_\mu u^\mu) u^\nu + u^\mu \partial_\mu u^\nu)\right] - u_\nu \partial^\nu P = \qquad (4.57)$$
$$= (u \cdot \partial)\epsilon + (\epsilon + P)\partial \cdot u,$$

where we have used the hydrodynamic parameterization of $T^{\mu\nu}$ in Eq. (2.61). Note that $u_\nu u^\mu \partial_\mu u^\nu = 0$ as a consequence of the normalization $u \cdot u = 1$. From this one has

$$\frac{\partial \epsilon}{\partial t} = -(\epsilon + P)\nabla \cdot \boldsymbol{u} = -\left(T\frac{\partial P}{\partial T} + \mu\frac{\partial P}{\partial \mu}\right)\nabla \cdot \boldsymbol{u}, \qquad (4.58)$$

where we have used Eq. (4.54). In a similar way using number density conservation one finds

$$\frac{\partial n}{\partial t} = -n\nabla \cdot \boldsymbol{u} = -\frac{\partial P}{\partial \mu}\nabla \cdot \boldsymbol{u}, \qquad (4.59)$$

where Eq. (4.55) has been used. The last two identities determine the right-hand side of Eq. (4.56) which is now written as

$$\frac{\mathrm{d}P(\epsilon,n)}{\mathrm{d}t} = -\left[\frac{\partial P}{\partial \epsilon}\left(T\frac{\partial P}{\partial T} + \mu\frac{\partial P}{\partial \mu}\right) + \frac{\partial P}{\partial n}\frac{\partial P}{\partial \mu}\right]\nabla \cdot \boldsymbol{u}. \qquad (4.60)$$

We are interested in the time-evolution of the thermal parameters T and μ. Comparing their coefficients in the left-hand side of Eq. (4.56) and Eq. (4.60) one finds:

$$\begin{aligned}\frac{\partial T}{\partial t} &= -T\frac{\partial P}{\partial \epsilon}\nabla \cdot \boldsymbol{u}, \\ \frac{\partial \mu}{\partial t} &= -\left[\mu\frac{\partial P}{\partial \epsilon} + \frac{\partial P}{\partial n}\right]\nabla \cdot \boldsymbol{u}.\end{aligned} \qquad (4.61)$$

The linear response of the energy-momentum tensor to the non-equilibrium state can then be evaluated and parameterized as[37]

$$\delta T^{ij} = \eta W^{ij} + \zeta \delta^{ij} \partial_k u^k, \qquad (4.62)$$

---

[37]Note that all entries containing a temporal coordinate simply vanish: $T^{\mu,0} = 0$. Of course, this is also true in the Kubo formalism.

again with a traceless tensor W^{ij} defined by

$$W^{ij} = \partial_i u^j + \partial_j u^i - \frac{2}{3}\delta_{ij}\partial_k u^k \,, \tag{4.63}$$

which should be compared to viscous hydrodynamics when parameterizing the dissipative tensor in Eq. (2.64). The final result for the shear viscosity reads [SR09]:

$$\begin{aligned}\eta(T,\mu) &= \beta \int \frac{\mathrm{d}^3 p}{(2\pi)^3} \frac{p_i^2 p_j^2}{2E^2} \left[\tau f_0^+ \left(1 \pm f_0^+\right) + \bar{\tau} f_0^- \left(1 \pm f_0^-\right)\right] = \\ &= \frac{\beta}{15} \int \frac{\mathrm{d}^3 p}{(2\pi)^3} \frac{p^4}{2E^2} \left[\tau f_0^+ \left(1 \pm f_0^+\right) + \bar{\tau} f_0^- \left(1 \pm f_0^-\right)\right], \end{aligned} \tag{4.64}$$

where the $\pm$ refers to the bosonic and fermionic case, respectively.

Let us finally compare the kinetic result for the shear viscosity with the functional $\eta[\Gamma]$ from Kubo formalism. As discussed in detail in Section 2.3.2, an infinite set of ladder diagrams must be resummed in order to get the full leading-order result for the shear viscosity in the weak-coupling limit. It was proven by Jeon that doing so the Kubo formalism is equivalent to kinetic theory [Jeo95]. Inspecting the results of this section we realize that using the Boltzmann equation in relaxation-time approximation one finds $\eta \sim \tau \sim 1/\Gamma$, which is indeed the same scaling as one finds from Kubo formalism at leading order in a weak-coupling expansion. The mathematical assumptions in both approaches, $\delta f \ll 1$ and $\langle B \rangle \ll \langle A \rangle$, just display the same physical picture that the considered system is close to equilibrium meaning that the relaxation time (or mean free time) is large. However, our result for $\eta[\Gamma]$ within the NJL model, Eq. (4.37), shows a more complex structure beyond the weak-coupling assumption. As we have discussed in Section 4.2.4, the spectral width Γ is far from being small, therefore the residual term A_{-1}/Γ in Eq. (4.43) is sub-dominant. Therefore ladder-diagram resummation within the NJL model is expected not to affect the numerical results for the shear viscosity as drastically as in weakly-coupled toy models or theories.

5. Mesonic fluctuations in the quark sector

> *"Quantum theory provides us with a striking illustration of the fact that we can fully understand a connection though we can only speak of it in images and parables."*[Hei71]
>
> Werner Heisenberg

In Section 3.3.2 we have discussed how a large-N_c analysis of the NJL model can reproduce standard techniques of many-body physics. The gap equation emerges at next-to-leading order, describing thermal constituent-quarks with dynamically generated masses, joined by and the Bethe-Salpeter equation (BSE) describing thermal mesons:

GAP: $\quad\blacksquare\quad = \quad\quad + \quad\bigcirc\quad + \quad\frown\quad$ (5.1)

BSE: $\quad\succ\!\!=\!\!\prec\quad = \quad\succ\!\!\prec\quad + \quad\succ\!\!\bigcirc\!\!\prec\quad$ (5.2)

The last diagram in the gap equation is of order $\mathcal{O}(N_c^{-1})$ and plays the role of the Fock term with respect to the leading-order gap equation at Hartree level. However, this term is necessary for a self-consistent treatment of both the quark and meson sector in a large-N_c NJL model. The Fock term introduces mesonic fluctuations into the quark sector and leads to a coupling between gap equation and the BSE. Therefore, the two equations have to be solved simultaneously which is not possible analytically. If we would be interested in thermodynamic properties of quarks and mesons only, a numerical approach would be sufficient. As we show in the first part of this chapter, producing discrete numerical data for the thermal quark self-energy is an insufficient starting point to derive its analytical continuation. This strategy is ill-defined and any predictive power for the shear viscosity is lost. Therefore, instead of solving numerically the two coupled equations (5.1) and (5.2) we first assume their decoupling and then use the solutions of the Hartree gap equation and the Bethe-Salpeter equation as input for the mesonic fluctuations.

5.1. Ambiguous analytical continuation from discrete data

Here we present a cautionary example of a discrete data set for which we try to find some analytical continuation for. It turns out that different ansatzes lead to dramatically different analytical properties of the final result. Therefore, no physical conclusion can be drawn solely on basis of this discrete data set. We consider the following model for some generic physical quantity $Q(p)$ in arbitrary units as function of momentum (with β denoting the inverse temperature as

5. Mesonic fluctuations in the quark sector

usual):

$$Q(p) = \frac{a_0}{1+\beta p} + \frac{a_1}{1+(\beta p)^2} + \sum_{k=2}^{6} a_k e^{-(\beta p)^k}, \qquad (5.3)$$

with $1 = 2a_0 = a_1 = a_3 = a_4 = a_5 = 2.5a_6$, and $\beta = (150 \text{ MeV})^{-1}$ which is arbitrary and chosen for convenience. We assume that we have access to just a discrete set of sample points $p_n \in \{0, 10, 20, \ldots, 350\}$ MeV, because, for instance, the underlying mathematical structure allows only for a numerical evaluation of Q.

The task now is to find an analytical function which describes the values of $Q(p)$ evaluated at the discrete values p_n without knowing the analytical form of $Q(p)$. For this, one might use different ansatzes, for instance the two following one:

$$\begin{aligned} Q_1(p) &= \frac{c_0}{1+(\beta p)^2} + c_2 e^{-c_3(\beta p)^2} + c_4 e^{-c_5(\beta p)^4} + c_6 e^{-c_7(\beta p)^6}, \\ Q_2(p) &= \frac{d_0}{1+\beta p} + d_1 e^{-d_2\beta p} + d_3 e^{-d_4(\beta p)^3} + d_5 e^{-d_6(\beta p)^5}, \end{aligned} \qquad (5.4)$$

with unknown coefficients c_i and d_i. Performing a least-squares fit leads to regression parameters given in Table 5.1. We realize that the comparison to the true coefficients defined by $Q(p)$ is rather unsatisfactory, but finally, both regressions meet the data set very well with an averaged relative error of 0.495% and 0.598%, relatively. In a graphical representation one can hardly see differences between the two regressions, compare Fig. 5.1. The difference between the two regressions is always less than 0.5% and the deviation from the true values of $Q(p)$ is in the worst case just 1.5%.

In Fig. 5.2 we show in addition, which momentum ranges contribute to the cumulated relative error between regression and the true data $Q(p)$. We conclude first that for momenta $p \lesssim 200$ MeV both regressions work better than for higher momenta. Second, there is actually no significant difference between the two regressions. Nevertheless, considering now their analytical continuations via $p_n \mapsto -ip$, we find two dramatically different results:

$$\text{Im}\, Q_1(-ip) = 0, \quad \text{Im}\, Q_2(-ip) \neq 0. \qquad (5.5)$$

Since the first ansatz, $Q_1(p)$, is an even function in p, its analytical continuation does not produce any imaginary part. In contrast, the ansatz $Q_2(p)$ contains also an odd part which implies $Q_2(-ip) \notin \mathbb{R}$.

In the Kubo formalism for transport coefficients imaginary parts of the quark self-energy govern the physics and are the crucial quantities to be derived. In a perturbative model one could simply expand the two-point function in Feynman diagrams deriving the self-energy. As discussed in the previous chapters, in the NJL model we apply a large-N_c expansion and describe

coefficient Q_1	c_0	c_2	c_3	c_4	c_5	c_6	c_7
fit value	1.94	0.81	1.58	2.20	0.85	0.93	1.17
true value	1	1	1	1	1	0.4	1
coefficient Q_2	d_0	d_1	d_2	d_3	d_4	d_5	d_6
fit value	−0.45	2.04	0.66	2.69	1.06	1.64	0.98
true value	0.5	0	∼	1	1	1	1

Table 5.1.: Fit results for the model Q_1 and Q_2 in comparison to their true analytical values. The tilde ∼ indicates that the fit parameter d_2 cannot be determined from $Q(p)$.

5.2. Quark self-energy from mesonic fluctuations

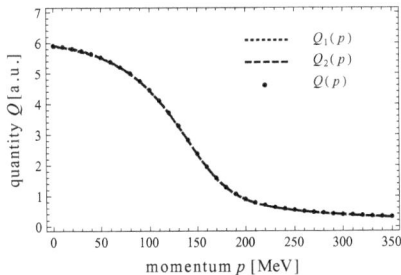

Figure 5.1.: Comparison between two least-squares fit (regressions) to the discrete data set generated by $Q(p)$. Graphically, there is no obvious difference between the two regressions, but their analytical properties differ dramatically, see the discussion in the text.

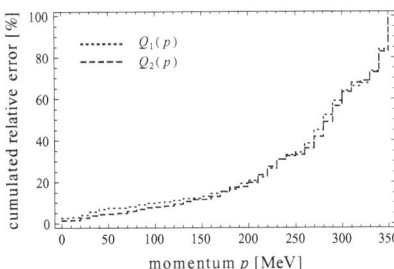

Figure 5.2.: Cumulated relative error for the two regressions $Q_1(p)$ and $Q_2(p)$. As in Fig. 5.2 there is no qualitative nor a big quantitative difference in the accuracy of the two regressions.

the quark-propagator by the self-consistent gap equation (5.1). From this the thermal spectral width can be extracted. We conclude that striving for just numerical solutions of this thermal self-energy leads to tremendous ambiguities for the transport coefficients, in particular for the shear viscosity which is governed by $\operatorname{Im} \Sigma_R(p)$. Our strategy is therefore to extract the imaginary part of the quark self-energy from mesonic fluctuations using thermal constituent-quark masses from the Hartree gap equation. This is a well-controlled procedure although it misses higher-order self-consistency corrections.

5.2. Quark self-energy from mesonic fluctuations

5.2.1. On-shell contributions

We now focus on the mesonic fluctuations which are described by the Fock term in the gap equation (5.1). They introduce a non-vanishing imaginary part of the quark self-energy at next-to-leading order in a large-N_c analysis. The Matsubara frequencies for a thermal constituent quark are $\nu_n = (2n+1)\pi T - i\mu$. Note that the frequencies for an antiquark read $\bar{\nu}_n = (2n+1)\pi T + i\mu = \nu_n^*$. There are $N_f^2 - 1$ contributing diagrams from the pseudoscalar channel (pions, $\Gamma^P = i\gamma_5 \tau_a$) and one diagram from the scalar channel (sigma boson, $\Gamma^S = \mathbb{1}$):

$$\Sigma_\beta^{S/P}(\boldsymbol{p}, \nu_n) = = \\ = g_{Mqq}^2 T \sum_{m \in \mathbb{Z}} \int \frac{d^3 q}{(2\pi)^3} \, \Gamma^{S/P} G_\beta^F(\boldsymbol{q}, \nu_m) \Gamma^{S/P} G_\beta^B(\boldsymbol{p}-\boldsymbol{q}, \nu_n - \nu_m) \,, \quad (5.6)$$

with the thermal quark and meson propagators, G_β^F and G_β^B, respectively, given in Appendix A.2. Due to its Dirac structure, there are three contributions to the thermal self-energy parameterized in the following form:

$$\Sigma_\beta^{S/P}(\boldsymbol{p}, \nu_n) = \pm m \Sigma_0 - \boldsymbol{p} \cdot \boldsymbol{\gamma} \Sigma_3 + \nu_n \gamma_4 \Sigma_4 \,, \quad (5.7)$$

5. Mesonic fluctuations in the quark sector

with three dimensionless functions Σ_i, for $i = 0, 3, 4$. The minus and plus sign in front of Σ_0 refers to the sigma boson and pion cases, respectively. They are given by

$$\Sigma_0(\boldsymbol{p}, \nu_n) = g_{\mathrm{Mqq}}^2 T \sum_{m \in \mathbb{Z}} \int \frac{\mathrm{d}^3 q}{(2\pi)^3} \frac{1}{\nu_m^2 + E_f^2} \frac{1}{(\nu_n - \nu_m)^2 + E_b^2},$$

$$\Sigma_3(\boldsymbol{p}, \nu_n) = g_{\mathrm{Mqq}}^2 T \sum_{m \in \mathbb{Z}} \int \frac{\mathrm{d}^3 q}{(2\pi)^3} \frac{\boldsymbol{p} \cdot \boldsymbol{q}}{p^2} \frac{1}{\nu_m^2 + E_f^2} \frac{1}{(\nu_n - \nu_m)^2 + E_b^2}, \quad (5.8)$$

$$\Sigma_4(\boldsymbol{p}, \nu_n) = g_{\mathrm{Mqq}}^2 T \sum_{m \in \mathbb{Z}} \int \frac{\mathrm{d}^3 q}{(2\pi)^3} \frac{\nu_m}{\nu_n} \frac{1}{\nu_m^2 + E_f^2} \frac{1}{(\nu_n - \nu_m)^2 + E_b^2},$$

with the energies $E_f^2 = q^2 + m^2$ and $E_b^2 = (\boldsymbol{q} - \boldsymbol{p})^2 + m_{\mathrm{M}}^2$. As always, the Matsubara sums can be carried out leading to some finite result with a combination of Bose and Fermi distribution functions. Technical details are shown in the Appendix, cf. Eqs. (A.24) and (A.25). We arrive at

$$\Sigma_{0,3}(\boldsymbol{p}, \nu_n) = g_{\mathrm{Mqq}}^2 \int \frac{\mathrm{d}^3 q}{(2\pi)^3} \mathcal{F}_{0,3} \left[\frac{1}{2 E_b E_f} \left(\frac{E_+ Z_1}{E_+^2 + \nu_n^2} + \frac{E_- Z_2}{E_-^2 + \nu_n^2} \right) + i\nu_n \frac{Z_3}{(E_+^2 + \nu_n^2)(E_-^2 + \nu_n^2)} \right], \quad (5.9)$$

with $E_\pm = E_b \pm E_f$. The quark-meson coupling, g_{Mqq}, can be pulled out of the integral since no momentum dependence is taken into account. This approximation has been discussed and justified in Section 3.5.2. We have introduced $\mathcal{F}_{0,3}$ as[38]

$$\mathcal{F}_0 = 1,$$
$$\mathcal{F}_3 = \frac{\boldsymbol{p} \cdot \boldsymbol{q}}{p^2} = \frac{m_{\mathrm{M}}^2 + p^2 + q^2 - E_b^2}{2 p^2}, \quad (5.10)$$

and have denoted the combinations of Bose and Fermi distributions as $Z_i(E_b, E_f)$:

$$Z_1 = 1 + n_{\mathrm{B}}(E_b) - \frac{1}{2}\left(n_{\mathrm{F}}^+(E_f) + n_{\mathrm{F}}^-(E_f)\right),$$
$$Z_2 = n_{\mathrm{B}}(E_b) + \frac{1}{2}\left(n_{\mathrm{F}}^+(E_f) + n_{\mathrm{F}}^-(E_f)\right) > 0, \quad (5.11)$$
$$Z_3 = n_{\mathrm{F}}^+(E_f) - n_{\mathrm{F}}^-(E_f) > 0.$$

The Bose and Fermi distributions read

$$n_{\mathrm{B}}(E) = \frac{1}{e^{\beta E} - 1}, \quad n_{\mathrm{F}}^\pm(E) = n_{\mathrm{F}}(E \mp \mu) = \frac{1}{e^{\beta(E \mp \mu)} + 1}, \quad (5.12)$$

where the $\pm$ signs denote quark and antiquark distribution functions, respectively. When carrying out the Matsubara sum also for the γ_4 part of the self-energy, we get:

$$\Sigma_4(\boldsymbol{p}, \nu_n) = g_{\mathrm{Mqq}}^2 \int \frac{\mathrm{d}^3 q}{(2\pi)^3} \left[\frac{1}{2 E_b} \left(\frac{Z_1}{E_+^2 + \nu_n^2} + \frac{Z_2}{E_-^2 + \nu_n^2} \right) - \frac{1}{2 i \nu_n} \frac{(E_b^2 - E_f^2 + \nu_n^2) Z_3}{(E_+^2 + \nu_n^2)(E_-^2 + \nu_n^2)} \right]. \quad (5.13)$$

In the next Section 5.3 we will discuss the vacuum limits of the self-energy contributions Σ_i, for the time being we focus on the calculation of their imaginary parts. They are crucial for evaluating the shear viscosity in Chapter 6. We start with investigating the new pole structure of the thermal quark-propagator at Fock level. Poles can only appear in Minkowski space, therefore we perform the analytical continuation via $\nu_n \mapsto -i p_0$. The mesonic fluctuations implies the

---
[38] Later when calculating imaginary parts of Σ_i, we will introduce also $\mathcal{F}_4$ in Eq. (5.35).

constituent-quark propagator:[39]

$$\frac{1}{\slashed{p} - m + \Sigma^{S/P}_\beta(\boldsymbol{p}, -ip_0)} = \frac{p_0\gamma_0(1+\Sigma_4) - \boldsymbol{p}\cdot\boldsymbol{\gamma}(1+\Sigma_3) + m(1\mp\Sigma_0)}{p_0^2(1+\Sigma_4)^2 - \boldsymbol{p}^2(1+\Sigma_3)^2 - m^2(1\mp\Sigma_0)^2}. \tag{5.14}$$

All Σ_i are evaluated in Minkowski space, e.g. the analytical continuation has been carried out: $(\boldsymbol{p}, \nu_n) \mapsto (\boldsymbol{p}, -ip_0 + \epsilon)$. We find at linear order in Σ_i the following pole condition:

$$p_0^2 = \boldsymbol{p}^2 + m^2 + \Omega^{S/P} \stackrel{!}{=} \boldsymbol{p}^2 + (m+\delta m)^2 \stackrel{!}{=} \left(\omega - \frac{i\widetilde{\Gamma}^{S/P}}{2}\right)^2, \tag{5.15}$$

with the *fermionic-pole correction*

$$\Omega^{S/P} = \boldsymbol{p}^2\left(2\Sigma_3 - 2\Sigma_4\right) + m^2\left(\mp 2\Sigma_0 - 2\Sigma_4\right). \tag{5.16}$$

We have denoted the leading-order quark energy by $\omega = \sqrt{\boldsymbol{p}^2 + m^2}$, and have introduced the (thermal) mass-shift, δm, and the resulting spectral width, $\widetilde{\Gamma}^{S/P}$. Neglecting again quadratic terms of the self-energy contributions Σ_i, we find

$$\begin{aligned}\delta m^{S/P} &= \frac{1}{2m}\operatorname{Re}\Omega^{S/P}, \\ \widetilde{\Gamma}^{S/P} &= -\frac{1}{\omega}\operatorname{Im}\Omega^{S/P}.\end{aligned} \tag{5.17}$$

Note that we have written the spectral width $\widetilde{\Gamma}$ with a tilde since it differs from the spectral width Γ we have used for the parameter study in Section 4.2 in Eq. (4.32). There we have assumed simplified quark self-energies with $\Sigma_3 = 0$ and $\Sigma_4 = 0$, ignoring the full Dirac structure of the quark propagator. Matching in this limit the pole structure in Eq. (5.14) with the ansatz for the quark propagator in Eq. (4.32), one finds:

$$\Gamma^{S/P} = -\frac{1}{2m}\operatorname{Im}\left(\mp\Sigma_0\right). \tag{5.18}$$

On the other hand one finds in this limit:

$$\widetilde{\Gamma}^{S/P} = -\frac{1}{\omega}\operatorname{Im}\Omega^{S/P} = -\frac{2m^2}{\omega}\operatorname{Im}\left(\mp\Sigma_0\right) = \frac{\omega}{2m}\Gamma^{S/P}. \tag{5.19}$$

As we will see later in Section 6.2, the shear viscosity is not just a function of *one* spectral width as defined in Eq. (5.17). The Dirac structure of the quark propagator induces a shear viscosity that indeed depends on the three imaginary parts of the self-energy contributions Σ_i and not only on a single spectral width. All details of the Kubo formalism including the full Dirac structure are discussed in Section 6.2.

The non-vanishing imaginary parts of Σ_i are induced by their pole structure:

$$\lim_{\epsilon\to 0}\operatorname{Im}\left.\frac{Z}{x^2 + \nu_n^2}\right|_{\nu_n\mapsto -ip_0+\epsilon} = Z\pi\delta(x^2 - p_0^2) = \frac{\pi Z}{2p_0}\left(\delta(x-p_0) + \delta(x+p_0)\right). \tag{5.20}$$

This means for the Z_1 term: $E_f + E_b \pm p_0 = 0$, where only the minus sign can be realized. For the Z_2 term, $E_f - E_b \pm p_0 = 0$, both signs can be realized for the time being. We will see that only the plus-sign case contributes to the (on-shell) imaginary parts, so there is just one contribution from Z_2. Later, the Z_3 term is considered separately. We start with the first two

---
[39]The denominator of the quark propagator actually reads $\slashed{p} - m - \Sigma_R(p_0, \boldsymbol{p})$. Following our sign convention, cf. Eq. (A.27) in the Appendix, the thermal self-energy enters with the opposite sign.

5. Mesonic fluctuations in the quark sector

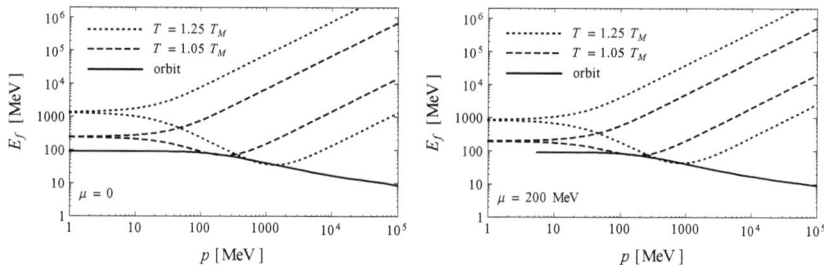

Figure 5.3.: Range of integration for $E_f \in [E_{\min}, E_{\max}]$ as function of momentum $p = |\mathbf{p}|$ for different temperatures $T = 1.05\,T_{\mathrm{M}}$ (dashed lines) and $T = 1.25\,T_{\mathrm{M}}$ (dotted lines). The solid line denotes the orbit of the minimal E_f when varying temperature, cf. Eq. (5.27).

terms $Z \in \{Z_1, Z_2\}$. Using the identify (5.20) we find the following structure when evaluating $\Sigma_{0,3}(\mathbf{p}, \nu_n)$ from Eq. (5.9) after analytical continuation has been carried out:

$$\int \frac{\mathrm{d}^3 q}{(2\pi)^3} \frac{\pi Z}{2p_0} \frac{1}{2E_b E_f} \delta(E_b - (*)) = \int \frac{\mathrm{d}^3 q}{(2\pi)^3} \frac{\pi Z}{2p_0 E_f} \delta(E_b^2 - (*)^2) =$$
$$= 2\pi \int_{-1}^{1} \mathrm{d}\xi \int_0^{\infty} \frac{\mathrm{d}q\, q^2}{(2\pi)^3} \frac{\pi Z}{2p_0 E_f} \delta(E_b^2(\xi) - (*)^2) = \quad (5.21)$$
$$= 2\pi \int_m^{\infty} \frac{\mathrm{d}E_f}{(2\pi)^3} \frac{\pi Z}{4p_0 |\mathbf{p}|} \Theta(1 - \xi^2),$$

where $\xi = \cos\theta$. In order to carry out the integral over the delta function we have used

$$E_b^2 = m_{\mathrm{M}}^2 + (\mathbf{p} - \mathbf{q})^2 = m_{\mathrm{M}}^2 + \mathbf{p}^2 + \mathbf{q}^2 - 2|\mathbf{p}||\mathbf{q}|\xi \quad \Rightarrow \quad \left|\frac{\partial E_b^2}{\partial \xi}\right| = 2|\mathbf{p}||\mathbf{q}|, \quad (5.22)$$

and converted the momentum integral to an energy integral using $|\mathbf{q}|\,\mathrm{d}q = E_f\,\mathrm{d}E_f$. The ill-conditioned Θ term can be removed by the following consideration: from Eq. (5.22) it is clear that $|\xi| \leq 1$ is fulfilled if and only if

$$-1 \leq \frac{E_b^2 - m_{\mathrm{M}}^2 - \mathbf{p}^2 - \mathbf{q}^2}{2|\mathbf{p}||\mathbf{q}|} \leq 1 \quad (5.23)$$
$$\Leftrightarrow F(E_f, \mathbf{p}) = 4\mathbf{p}^2(E_f^2 - m^2) - \left[E_b^2 - m_{\mathrm{M}}^2 - \mathbf{p}^2 + m^2 - E_f^2\right]^2 \geq 0.$$

For a given value of the absolute momentum the roots of $F(\cdot, \mathbf{p})$ read for the plus-sign case $0 = E_f + E_b + p_0$, and therefore $E_b^2 = (E_f + p_0)^2$:

$$E_{\max,\min} = \frac{1}{2m^2}\left[(m_{\mathrm{M}}^2 - 2m^2)\sqrt{m^2 + \mathbf{p}^2} \pm |\mathbf{p}|m_{\mathrm{M}}\sqrt{m_{\mathrm{M}}^2 - 4m^2}\right] \quad (5.24)$$

The range if integration, $E_f \in [E_{\min}, E_{\max}]$, depends therefore linearly on the external quark momentum:

$$E_{\max} - E_{\min} = \frac{|\mathbf{p}|m_{\mathrm{M}}}{m^2}\sqrt{m_{\mathrm{M}}^2 - 4m^2}. \quad (5.25)$$

In the limit of a vanishing external quark momentum the range of integration collapses to one

single point:
$$E_{\text{max,min}}|_{|\boldsymbol{p}|=0} = \frac{m_M^2}{2m} - m > m \, . \tag{5.26}$$

We emphasize that the whole discussion is only valid for temperatures above the Mott temperatures T_M, where the pion mass is at least twice the constituent-quark mass. This constraint can be seen explicitly from Eq. (5.24). We have already introduced the Mott temperature when discussing thermal quark and meson masses in Section 3.5, where $T_M \approx 212$ MeV have been found in the case of vanishing quark chemical potential. Note that this discussion remains valid also in the chiral limit, where the current-quark mass is set to zero, $m_0 = 0$. In this case, the pion mass vanishes in the Nambu-Goldstone phase at low temperatures but it is finite when chiral symmetry is restored for large temperatures, cf. the NJL phase diagram shown in Fig. 3.5.

Fig. 5.3 shows the momentum-dependent phase space for E_f for different temperatures including all thermal effects for quarks, $m(T,\mu)$, and mesons, $m_M(T,\mu)$. Due to the large meson mass at high T, the curves $E_{\text{max,min}}$ are shifted to higher energies and momenta when increasing the temperature. The minimal value of $E_{\min}(p)$ is always m (in agreement with $E_f \geq m$) and is reached at
$$p_{\text{minimizer}} = \frac{m_M}{2m}\sqrt{m_M^2 - 4m^2} \, . \tag{5.27}$$

The upper boundary of the range of integration is a monotonic function of momentum and reaches its minimal value at $p = 0$.

We conclude that under the condition $m_M > 2m$, i.e. for $T > T_M$, the phase space is always non-empty and compact: $\emptyset \neq [E_{\min}, E_{\max}] \subseteq [m, \infty)$. This fact implies that the shear viscosity η will evaluate to some finite result in this temperature region. However, we have also derived the following substitution rule
$$\int_m^\infty \frac{\mathrm{d}E_f}{(2\pi)^3}(\,\cdot\,)\Theta(1-\xi^2) = \int_{E_{\min}}^{E_{\max}} \frac{\mathrm{d}E_f}{(2\pi)^3}(\,\cdot\,) \, , \tag{5.28}$$

which leads finally to a well-conditioned one-dimensional numerical integral.

For the sake of completeness, we also mention the minus-sign case, i.e. $0 = E_b + E_f - p_0$. If we plug in $E_b^2 = (E_f - p_0)^2$ into the condition (5.23) then the phase space simply vanishes for any incoming quark momentum, since the range of integration would be restricted to negative energies in the fermion loop:
$$E'_{\min} = -E_{\max}, \quad E'_{\max} = -E_{\min} \, . \tag{5.29}$$

We can therefore conclude that only the plus-sign case, $E_b = E_f + p_0$, allows for an on-shell condition for the mesonic fluctuation. Knowing this we can now continue with the third term, Z_3, in Eq. (5.9):
$$\begin{aligned}\lim_{\epsilon \to 0} \text{Im}\, \frac{i\nu_n Z_3}{[(E_f + E_b)^2 + \nu_n^2][(E_f - E_b)^2 + \nu_n^2]}\bigg|_{i\nu_n \mapsto p_0 + i\epsilon} &= \\ = p_0 \pi Z_3\, \delta\left([(E_f + E_b)^2 - p_0^2][(E_f - E_b)^2 - p_0^2]\right) &= \\ = \frac{p_0 \pi Z_3}{2p_0}\, \delta\big(\underbrace{[(E_f + E_b)^2 - p_0^2]}_{=4E_f E_b}[E_b - E_f - p_0]\big) &= \\ = \frac{\pi Z_3}{4E_f}\, \delta\left(E_b^2 - (E_f + p_0)^2\right) . \end{aligned} \tag{5.30}$$

Note that due to the $i\nu_n$ factor in the first line, the p_0 terms cancel in the final result. As done

5. Mesonic fluctuations in the quark sector

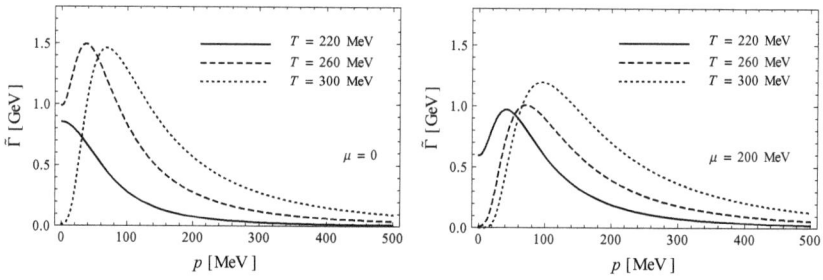

Figure 5.4.: Summed spectral width $\tilde{\Gamma}$ as function of momentum $p = |\boldsymbol{p}|$ for different temperatures T and for vanishing quark chemical potential (left) and $\mu = 200$ MeV (right)

in the calculation (5.21) the momentum integral can be performed:

$$\int \frac{\mathrm{d}^3 q}{(2\pi)^3} \frac{\pi Z_3}{4 E_f} \delta(E_b^2 - (E_f + p_0)^2) = 2\pi \int_m^\infty \frac{\mathrm{d}E_f}{(2\pi)^3} \frac{\pi Z_3}{8|\boldsymbol{p}|} \Theta(1 - \xi^2) \,. \tag{5.31}$$

Combining all contributions, we find for the imaginary parts of Σ_0 and Σ_3:

$$\begin{aligned}
\mathrm{Im}\,\Sigma_{0,3}(\boldsymbol{p}, -\mathrm{i}p_0) &= 2\pi g_{\mathrm{Mqq}}^2 \int_{E_{\min}}^{E_{\max}} \frac{\mathrm{d}E_f}{(2\pi)^3} \mathcal{F}_{0,3} \frac{\pi}{4|\boldsymbol{p}|} \left(\frac{(E_f - E_b)Z_2}{p_0} + \frac{Z_3}{2} \right) = \\
&= -\frac{g_{\mathrm{Mqq}}^2}{16\pi|\boldsymbol{p}|} \int_{E_{\min}}^{E_{\max}} \mathrm{d}E_f\, \mathcal{F}_{0,3} \left(Z_2 - \frac{Z_3}{2} \right) = \\
&= -\frac{g_{\mathrm{Mqq}}^2}{16\pi|\boldsymbol{p}|} \int_{E_{\min}}^{E_{\max}} \mathrm{d}E_f\, \mathcal{F}_{0,3} \left[n_{\mathrm{B}}(E_b) + n_{\mathrm{F}}^-(E_f) \right],
\end{aligned} \tag{5.32}$$

using $E_b = E_f + p_0$ from the on-shell condition. It now remains to calculate the imaginary part of Σ_4 as well. Due to the identical pole structure, this result can be easily obtained by simply adjusting the prefactors and imitating the calculation as it has been done for $\Sigma_{0,3}$. One gets:

$$\begin{aligned}
\mathrm{Im}\,\Sigma_4(\boldsymbol{p}, -\mathrm{i}p_0) &= 2\pi g_{\mathrm{Mqq}}^2 \int_{E_{\min}}^{E_{\max}} \frac{\mathrm{d}E_f}{(2\pi)^3} \frac{\pi}{4|\boldsymbol{p}|} \left(\frac{E_f Z_2}{p_0} - \frac{(E_b^2 - E_f^2 - p_0^2)Z_3}{4 p_0^2} \right) = \\
&= \frac{g_{\mathrm{Mqq}}^2}{16\pi|\boldsymbol{p}|} \int_{E_{\min}}^{E_{\max}} \mathrm{d}E_f\, \frac{E_f}{p_0} \left[n_{\mathrm{B}}(E_b) + n_{\mathrm{F}}^-(E_f) \right],
\end{aligned} \tag{5.33}$$

using again the on-shell condition resulting in $E_b^2 - E_f^2 - p_0^2 = 2 p_0 E_f$.

In summary, the imaginary parts of the three Dirac components of the quark-self energy $\Sigma_\beta(\boldsymbol{p}, \nu_n)$ in Eq. (5.7) read

$$\mathrm{Im}\,\Sigma_{0,3,4}(\boldsymbol{p}, -\mathrm{i}p_0) = -\frac{g_{\mathrm{Mqq}}^2}{16\pi|\boldsymbol{p}|} \int_{E_{\min}}^{E_{\max}} \mathrm{d}E_f\, \mathcal{F}_{0,3,4} \left[n_{\mathrm{B}}(E_b) + n_{\mathrm{F}}^-(E_f) \right], \tag{5.34}$$

with $\mathcal{F}_{0,3}$ defined in Eq. (5.10) and

$$\mathcal{F}_4 = -\frac{E_f}{p_0}\,. \tag{5.35}$$

With these results we are now able to determine the spectral width $\tilde{\Gamma}^{\mathrm{S/P}}$ as defined in Eq. (5.17):

5.2. Quark self-energy from mesonic fluctuations

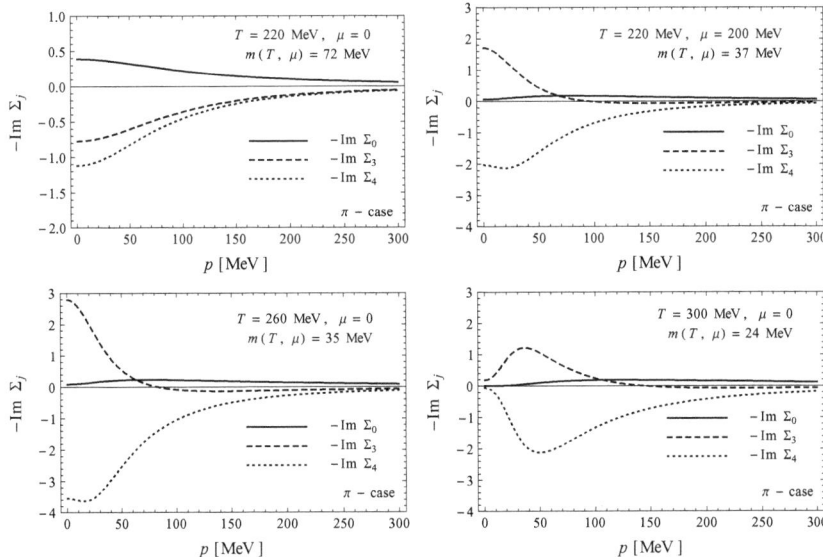

Figure 5.5.: The (negative) imaginary part of the self-energy contributions Σ_j, $j = 0, 3, 4$ from mesonic fluctuations. They have been defined in Eq. (5.7). See also the discussion in the text.

$$\widetilde{\Gamma}^{\mathrm{S/P}}(\boldsymbol{p}, -\mathrm{i}p_0) = -\frac{1}{p_0}\mathrm{Im}\,\Omega^{\mathrm{S/P}} = \frac{g_{\mathrm{Mqq}}^2 m_\pi^2}{16\pi p_0 |\boldsymbol{p}|} \int_{E_{\min}}^{E_{\max}} \mathrm{d}E_f \left[n_{\mathrm{B}}(E_b) + n_{\mathrm{F}}^-(E_f)\right] =$$
$$= \frac{g_{\mathrm{Mqq}}^2 m_\pi^2}{16\pi p_0 |\boldsymbol{p}|} T \ln \frac{n_{\mathrm{F}}^-(E_{\max})\, n_{\mathrm{B}}(E_{\min} + p_0)}{n_{\mathrm{F}}^-(E_{\min})\, n_{\mathrm{B}}(E_{\max} + p_0)} . \quad (5.36)$$

Both scalar and pseudoscalar channels are described by the same functional structure and, in particular, with the same prefactor m_π^2. The different meson masses affect only the numerical value of the boundaries, $E_{\min/\max}$, defined in Eq. (5.24). This remarkable feature is due to

$$\mathrm{Im}\,\Omega^{\mathrm{S/P}} \sim \begin{cases} \boldsymbol{p}^2(\mathcal{F}_3 - \mathcal{F}_4) + m^2(-\mathcal{F}_0 - \mathcal{F}_4) = \tfrac{1}{2}\left(m_{\mathrm{M}}^2 - 4m^2\right) = \tfrac{1}{2}m_\pi^2 & \text{for the } \sigma\text{-case}, \\ \boldsymbol{p}^2(\mathcal{F}_3 - \mathcal{F}_4) + m^2(+\mathcal{F}_0 - \mathcal{F}_4) = \tfrac{1}{2}m_{\mathrm{M}}^2 = \tfrac{1}{2}m_\pi^2 & \text{for the } \pi\text{-case}, \end{cases} \quad (5.37)$$

where we have used the relation (3.102) between pion and sigma-meson mass. In Fig. 5.4 we show the momentum dependence of the summed spectral width $\widetilde{\Gamma} = 3\widetilde{\Gamma}^{\mathrm{P}} + \widetilde{\Gamma}^{\mathrm{S}}$, at different values for temperature and quark chemical potential. We recall that $\widetilde{\Gamma}$ denotes the imaginary part of the pole condition of the quark propagator including mesonic fluctuations at order $1/N_{\mathrm{c}}$, cf. Eq. (5.16) and the related discussion. This effectively combined spectral width turns out to be at the order of 1 GeV, which is huge compared to typical NJL-model scales as its low-energy spectrum represented by $m_\pi^{\mathrm{vac}} = 140$ MeV, the constituent-quark mass scale $m^{\mathrm{vac}} = 325$ MeV, or its ultraviolet cutoff $\Lambda = 651$ MeV. However, as already mentioned, this unphysically large width does not affect the shear viscosity directly. As we will describe in Section 6.2, the shear viscosity actually depends on the three independent imaginary parts of Σ_j ($j = 0, 3, 4$) rather

5. Mesonic fluctuations in the quark sector

than on the pole's imaginary part $\tilde{\Gamma}$. They are shown in Fig. 5.5, again for different values of temperature and chemical potential. From Eq. (5.7) these quantities have been defined to be dimensionless, hence a comparison with $\tilde{\Gamma}$ is possible only when introducing some mass scale[40]. In comparison to $\tilde{\Gamma}$, the overall scale of all Im Σ_j is much smaller but also qualitative differences occur: Only $-\text{Im } \Sigma_0$ stays positive and can be interpreted as spectral width by itself. In contrast, Im Σ_3 and Im Σ_4 can be negative as well, depending on the triple (p, T, μ). We have chosen to show the imaginary parts for the pion case, since results for the sigma boson are both qualitatively and quantitatively (almost) the same. These imaginary part determine the shear viscosity in a non-trivial way (cf. Eq. (6.26)) as it will be discussed in Section 6.2.

For the numerical results of Im Σ_j for $j = 0, 3, 4$ we have performed the remaining energy integration, dE_f in Eq. (5.34), which allows to express all imaginary parts as analytical functions. Starting with Im Σ_0, we have due to $\mathcal{F}_0 = 1$ the same integral as carried out before for $\tilde{\Gamma}^{S/P}$:

$$\text{Im } \Sigma_0(\bm{p}, -ip_0) = \frac{g_{\text{Mqq}}^2}{16\pi|\bm{p}|} T \ln \frac{n_{\text{F}}^-(E_{\min})\, n_{\text{B}}(E_{\max} + p_0)}{n_{\text{F}}^-(E_{\max})\, n_{\text{B}}(E_{\min} + p_0)}. \tag{5.38}$$

The Dirac parts Σ_3 and Σ_4 contain some energy-dependent prefactor, $\mathcal{F}_3$ and $\mathcal{F}_4$, respectively, leading to some more complex final result. We introduce the auxiliary function

$$\mathcal{H}(E) = (E + p_0) \ln n_{\text{F}}^-(E) - T \text{Li}_2\left(-\frac{1}{n_{\text{B}}(E + p_0)}\right) - T \text{Li}_2\left(1 - \frac{1}{n_{\text{F}}^-(E)}\right). \tag{5.39}$$

It follows that:

$$\text{Im } \Sigma_3(\bm{p}, -ip_0) = \left(1 + \frac{m_{\text{M}}^2}{2|\bm{p}|^2}\right) \text{Im } \Sigma_0 + \frac{g_{\text{Mqq}}^2 p_0}{16\pi|\bm{p}|^3} T \mathcal{H}(E)\big|_{E_{\min}}^{E_{\max}},$$

$$\text{Im } \Sigma_4(\bm{p}, -ip_0) = \text{Im } \Sigma_0 + \frac{g_{\text{Mqq}}^2}{16\pi|\bm{p}|p_0} T \mathcal{H}(E)\big|_{E_{\min}}^{E_{\max}}. \tag{5.40}$$

These results for Im $\Sigma_{0,3,4}$ will be used for the evaluation of the shear viscosity (6.26) in Chapter 6. Due to their analytical structure, they can be handled easily, therefore, numerical issues arise solely from the peak structure of the underlying Kubo formula.

5.2.2. Off-shell contributions

So far we have treated the external quark in the Fock diagram $\Sigma_\beta^{S/P}(\bm{p}, -ip_0)$ in Eq. (5.6) as an on-shell particle, i.e. when determining the imaginary parts of $\Sigma_{0,3,4}(\bm{p}, -ip_0)$ we have used the dispersion relation $p_0^2 = \bm{p}^2 + m^2$. Inspecting the general Kubo formula for the shear viscosity derived in Eq. (4.30), we recall that the spectral function $\rho(p_0, \bm{p})$ in its integrand is generally defined off-shell. Using on-shell expressions for Im $\Sigma_{0,3,4}(\bm{p}, -ip_0)$ is a convenient but unnecessary approximation. In the following we will derive analytical results for the off-shell imaginary parts of the quark self-energy from mesonic fluctuations. We return to $\Sigma_{0,3}$ given in Eq. (5.9) and decompose into partial fractions for convenience:

$$\Sigma_{0,3}(\bm{p}, -ip_0) = g_{\text{Mqq}}^2 \int \frac{d^3q}{(2\pi)^3} \frac{\mathcal{F}_{0,3}}{4 E_b E_f} \left[\frac{1 - n_{\text{F}}^-(E_f) + n_{\text{B}}(E_b)}{E_f + E_b + p_0} + \right.$$
$$\left. + \frac{n_{\text{B}}(E_b) + n_{\text{F}}^+(E_f)}{E_f - E_b + p_0 + i\epsilon} + \frac{n_{\text{B}}(E_b) + n_{\text{F}}^-(E_f)}{E_f - E_b - p_0 - i\epsilon} + \frac{1 + n_{\text{B}}(E_b) - n_{\text{F}}^+(E_f)}{E_f + E_b - p_0 - i\epsilon} \right]. \tag{5.41}$$

---
[40] As it can be seen in Eq. (5.7), Σ_0 is multiplied by m. Hence, it is natural to use the thermal constituent-quark mass to set the scale as it is done in Fig. 5.5.

5.2. Quark self-energy from mesonic fluctuations

As in the on-shell discussion, taking its imaginary part probes the pole position of the partial fractions introducing four cases $\pm E_b = E_f \pm p_0$. The fraction in the first line introduces $E_b = -E_f - p_0 < 0$ which can be excluded immediately. The remaining three cases are denoted as follows:

$$\begin{aligned} \text{Case} \quad &\text{I:} \quad E_b = E_f + p_0 \,, \\ \text{Case} \quad &\text{II:} \quad E_b = E_f - p_0 \,, \\ \text{Case} \quad &\text{III:} \quad E_b = p_0 - E_f \,. \end{aligned} \quad (5.42)$$

Carrying out the $\mathrm{d}^3 q$ integral introduces again the restriction $|\xi| \leq 1$ with $\xi = \cos\theta$ denoting the polar angle, c.f. Eq. (5.23):

$$-1 \leq \frac{E_b(|\boldsymbol{p}|, p_0)^2 - m_\mathrm{M}^2 - \boldsymbol{p}^2 - \boldsymbol{q}^2}{2|\boldsymbol{p}||\boldsymbol{q}|} \leq 1 \quad (5.43)$$
$$\Leftrightarrow \quad F(E_f, |\boldsymbol{p}|, p_0) = 4\boldsymbol{p}^2(E_f^2 - m^2) - \left[E_b(|\boldsymbol{p}|, p_0)^2 - m_\mathrm{M}^2 - \boldsymbol{p}^2 + m^2 - E_f^2 \right]^2 \geq 0 \,,$$

but in the off-shell case $|\boldsymbol{p}| > 0$ and $p_0 \geq m$ are independent of each other. In the following we evaluate the three-dimensional integral (5.41) ensuring $|\xi| \leq 1$ by applying the three cases for the relation between quark and meson energy. We start with

Case I. This is the only case that can be realized on-shell: $E_b = E_f + p_0$. The following two conditions have to be fulfilled: (i) $E_f > m$ and (ii) $E_f > m_\mathrm{M} - p_0$, which can be summarized in $E_f > \max(m, m_\mathrm{M} - p_0)$. Evaluating the condition $|\xi| \leq 1$ we find

$$F(E_f, |\boldsymbol{p}|, p_0) \geq 0 \quad \Leftrightarrow \quad -4s(E_f - \widetilde{E}_-)(E_f - \widetilde{E}_+) \geq 0 \,, \quad (5.44)$$

where we have introduced $s = p_0^2 - \boldsymbol{p}^2$ and

$$\widetilde{E}_\pm = -\frac{p_0}{2} + \frac{(m_\mathrm{M}^2 - m^2)p_0}{2s} \pm \frac{|\boldsymbol{p}|}{2s}\sqrt{[s - (m+m_\mathrm{M})^2][s - (m-m_\mathrm{M})^2]} \,. \quad (5.45)$$

These roots of $F(E_f, \cdot, \cdot)$ are generalizations of $E_{\max,\min}$ introduced in Eq. (5.24). One finds indeed

$$\widetilde{E}_\pm \Big|_{s = p_0^2 - \boldsymbol{p}^2 = m^2 > 0} = E_{\max,\min} \,. \quad (5.46)$$

Note that in contrast to $m < E_{\min} < E_{\max}$, the off-shell roots are not ordered that simply. Dependent on $s > 0$ or $s < 0$ one has $\widetilde{E}_- < \widetilde{E}_+$ or $\widetilde{E}_+ < \widetilde{E}_-$, respectively. In addition, it might happen that one or even both roots are negative as we will see.

First, we consider the case $s < 0$ which leads to a convex-up parabola $F(E_f, \cdot, \cdot)$ with possible integration ranges $E_f < \widetilde{E}_+$ and $E_f > \widetilde{E}_-$. In general one has to distinguish additionally the two cases $m < m_\mathrm{M}$ and $m > m_\mathrm{M}$, but right now we find for both cases

$$\widetilde{E}_+ < -m_\mathrm{M} - p_0 < 0 \,, \quad \widetilde{E}_- > \max(m, m_{\mathrm{M}-p_0}) \,. \quad (5.47)$$

For Case I with $s < 0$ we have the range of integration $E_f > \widetilde{E}_-$ as sketched in Fig. 5.6(a).

Now consider the case $s > 0$ with a concave-down parabola $F(E_f, \cdot, \cdot)$. The possible integration range is $\widetilde{E}_- < E_f < \widetilde{E}_+$. This time, the roots are not automatically real numbers, but for $(m - m_\mathrm{M})^2 < s < (m + m_\mathrm{M})^2$ they become purely imaginary and have to be excluded. The first option $p_0 < \sqrt{(m - m_\mathrm{M})^2 + \boldsymbol{p}^2}$ leads to

$$\begin{aligned} m < m_\mathrm{M}: \quad &\widetilde{E}_+ > \widetilde{E}_- > \max(m, m_\mathrm{M} - p_0) \,, \\ m > m_\mathrm{M}: \quad &\widetilde{E}_- < \widetilde{E}_+ < -m - p_0 < 0 \,. \end{aligned} \quad (5.48)$$

5. Mesonic fluctuations in the quark sector

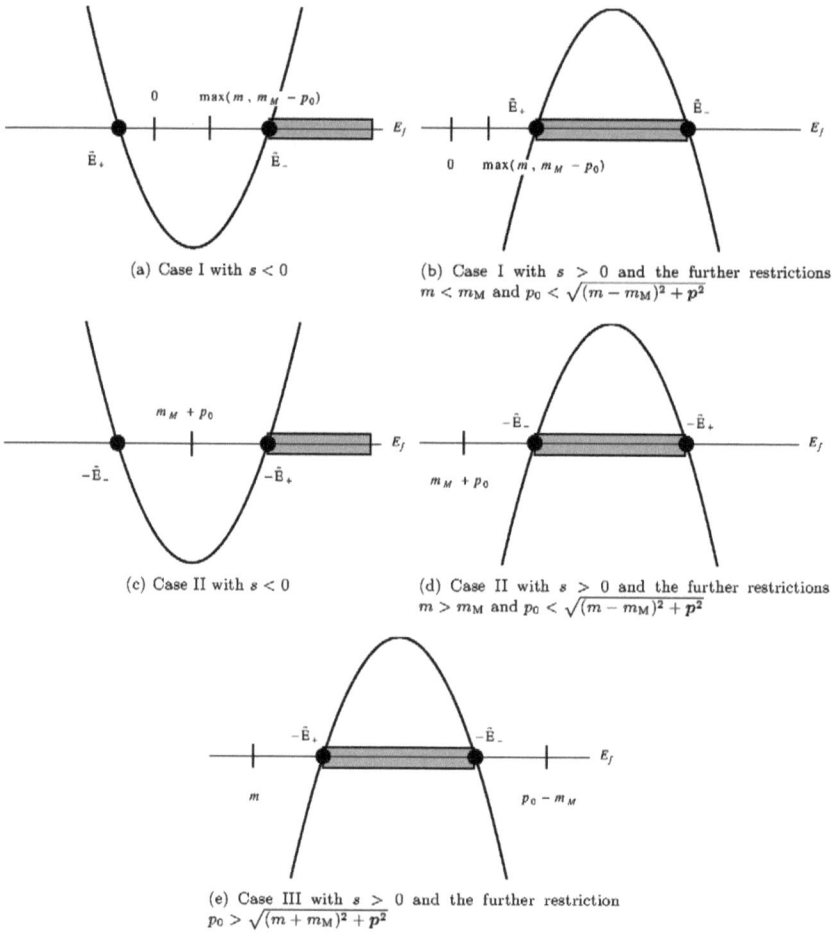

Figure 5.6.: Summary of integration ranges (gray boxes) for the off-shell imaginary parts of $\Sigma_{0,3,4}(\boldsymbol{p}, -ip_0)$. On-shell only the case (b) can be realized. See the discussion in the text.

Therefore, the case $m > m_M$ cannot be realized and only for $m < m_M$ the full range of integration is accessible. We summarize this case in Fig. 5.6(b). Having $s > 0$ there is the second option $p_0 > \sqrt{(m + m_M)^2 + \boldsymbol{p}^2}$ for which one has for both cases $m < m_M$ and $m > m_M$:

$$\tilde{E}_+ < -m, \quad \tilde{E}_- > m_M - p_0. \tag{5.49}$$

We conclude $m_M - p_0 < \tilde{E}_- < \tilde{E}_+ < -m < 0$, hence this option is excluded and the discussion of Case I is completed.

Case II. Evaluating the condition $|\xi| \leq 1$ using $E_b = E_f - p_0$ leads to

$$F(E_f, p, p_0) \geq 0 \quad -4s(E_f + \widetilde{E}_-)(E_f + \widetilde{E}_+) \geq 0, \quad (5.50)$$

hence $-\widetilde{E}_\pm$ are the roots of $F(E_f, \cdot, \cdot)$. We follow the same systematic path as before:

Consider first the case $s < 0$ implying again a convex-up parabola with $\widetilde{E}_- > \widetilde{E}_+$. This means $-\widetilde{E}_- < -\widetilde{E}_+$, providing two possible integration ranges $E_f < -\widetilde{E}_-$ and $E_f > -\widetilde{E}_+$. One finds:

$$\begin{aligned} m < m_{\mathrm{M}}: &\quad -\widetilde{E}_+ > m_{\mathrm{M}} + p_0, \quad -\widetilde{E}_- < -m_{\mathrm{M}} + p_0 < m_{\mathrm{M}} + p_0, \\ m > m_{\mathrm{M}}: &\quad -\widetilde{E}_+ > m_{\mathrm{M}} + p_0, \quad -\widetilde{E}_- < -m + p_0 < -m_{\mathrm{M}} + p_0 < m_{\mathrm{M}} + p_0. \end{aligned} \quad (5.51)$$

In conclusion we find the range of integration as shown in Fig. 5.6(c), again without any restriction on the quark and meson masses.

Now consider $s > 0$. This time the possible range of integration is $-\widetilde{E}_+ < E_f < -\widetilde{E}_-$. For the option $p_0 < \sqrt{(m - m_{\mathrm{M}})^2 + \boldsymbol{p}^2}$ we find

$$\begin{aligned} m < m_{\mathrm{M}}: &\quad -\widetilde{E}_+ < -\widetilde{E}_- < -m_{\mathrm{M}} + p_0 < m_{\mathrm{M}} + p_0, \\ m > m_{\mathrm{M}}: &\quad -\widetilde{E}_- > -\widetilde{E}_+ > m_{\mathrm{M}} + p_0. \end{aligned} \quad (5.52)$$

Using the constraint $E_f > m_{\mathrm{M}} + p_0$ the case $m < m_{\mathrm{M}}$ is excluded and only $m > m_{\mathrm{M}}$ is possible. The option $p_0 > \sqrt{(m + m_{\mathrm{M}})^2 + \boldsymbol{p}^2}$, for both cases $m < m_{\mathrm{M}}$ and $m > m_{\mathrm{M}}$, leads to:

$$-\widetilde{E}_+ < -\widetilde{E}_- < -m_{\mathrm{M}} + p_0 < m_{\mathrm{M}} + p_0, \quad (5.53)$$

which excludes this case because $E_f > \max(m, m_{\mathrm{M}} + p_0)$ must be ensured. This case is illustrated in Fig. 5.6(d).

Case III. The final case, $E_b = p_0 - E_f$, leads to the two conditions (i) $E_f > m$ and (ii) $E_f < p_0 - m_{\mathrm{M}}$. From this we get $p_0 > m + m_{\mathrm{M}}$. It is important to realize that E_b in this case is just the negative of the condition used in Case I. Therefore, all contributions present for Case I cannot be realized for Case III. It remains to check the case $s > 0$ in combination with $p_0 > \sqrt{(m + m_{\mathrm{M}})^2 + \boldsymbol{p}^2}$. We have $-\widetilde{E}_+ < E_f < -\widetilde{E}_-$ as possible integration range and find (cf. the related discussion for Case I):

$$-\widetilde{E}_+ > m, \quad -\widetilde{E}_- < p_0 - m_{\mathrm{M}}, \quad (5.54)$$

which is valid for both $m < m_{\mathrm{M}}$ and $m > m_{\mathrm{M}}$. In conclusion, there is only one contribution to the imaginary part for Case III as shown in Fig. 5.6(e).

Combining now all three cases, the off-shell imaginary part of $\Sigma_0(\boldsymbol{p}, -ip_0)$ can be calculated immediately. The rather lengthly result reads[41]

$$\begin{aligned} \mathrm{Im}\,\Sigma_0^{\mathrm{off}} = \frac{g_{\mathrm{Mqq}}^2}{16\pi|\boldsymbol{p}|} \bigg\{ &\int_{\mathrm{I}} \mathrm{d}E_f \left[-n_{\mathrm{B}}(E_f + p_0) - n_{\mathrm{F}}^-(E_f)\right] + \\ &+ \int_{\mathrm{II}} \mathrm{d}E_f \left[n_{\mathrm{B}}(E_f - p_0) + n_{\mathrm{F}}^+(E_f)\right] + \\ &+ \int_{\mathrm{III}} \mathrm{d}E_f \left[1 + n_{\mathrm{B}}(p_0 - E_f) - n_{\mathrm{F}}^+(E_f)\right] \bigg\} = \frac{g_{\mathrm{Mqq}}^2}{16\pi|\boldsymbol{p}|} \left(J^{\mathrm{I}} + J^{\mathrm{II}} + J^{\mathrm{III}}\right), \end{aligned} \quad (5.55)$$

---

[41] The minus signs for Case I is due to the pole description $+i\epsilon$ instead of $-i\epsilon$ for Case II and Case III. For E_b we have always inserted the corresponding relations to E_f and p_0 as defined in Eq. (5.42).

5. Mesonic fluctuations in the quark sector

with

$$
\begin{aligned}
J^{\mathrm{I}} = {}& \theta(|\boldsymbol{p}|-p_0)\left[\mu - p_0 + T\ln\frac{n_{\mathrm{F}}^-(\widetilde{E}_-)}{n_{\mathrm{B}}(\widetilde{E}_- + p_0)}\right] + \\
& + \theta(p_0 - |\boldsymbol{p}|)\theta(m_{\mathrm{M}} - m)\theta(\sqrt{(m-m_{\mathrm{M}})^2 + \boldsymbol{p}^2} - p_0)\, T\ln\frac{n_{\mathrm{F}}^-(\widetilde{E}_-)\,n_{\mathrm{B}}(\widetilde{E}_+ + p_0)}{n_{\mathrm{F}}^-(\widetilde{E}_+)\,n_{\mathrm{B}}(\widetilde{E}_- + p_0)},
\end{aligned} \tag{5.56}
$$

$$
\begin{aligned}
J^{\mathrm{II}} = {}& \theta(|\boldsymbol{p}|-p_0)\left[\mu - p_0 + T\ln\frac{n_{\mathrm{B}}(-\widetilde{E}_+ - p_0)}{n_{\mathrm{F}}^+(-\widetilde{E}_+)}\right] + \\
& + \theta(p_0 - |\boldsymbol{p}|)\theta(m - m_{\mathrm{M}})\theta(\sqrt{(m-m_{\mathrm{M}})^2 + \boldsymbol{p}^2} - p_0)\, T\ln\frac{n_{\mathrm{F}}^+(-\widetilde{E}_-)\,n_{\mathrm{B}}(-\widetilde{E}_+ - p_0)}{n_{\mathrm{F}}^+(-\widetilde{E}_+)\,n_{\mathrm{B}}(-\widetilde{E}_- - p_0)},
\end{aligned} \tag{5.57}
$$

$$
J^{\mathrm{III}} = \theta(p_0 - \sqrt{(m+m_{\mathrm{M}})^2 + \boldsymbol{p}^2})\, T\ln\frac{n_{\mathrm{F}}^-(\widetilde{E}_+)\,n_{\mathrm{B}}(\widetilde{E}_- + p_0)}{n_{\mathrm{F}}^-(\widetilde{E}_-)\,n_{\mathrm{B}}(\widetilde{E}_+ + p_0)}. \tag{5.58}
$$

Note that for J^{III} the condition $\theta(p_0 - |\boldsymbol{p}|)$ just follows from $\theta(p_0 - \sqrt{(m+m_{\mathrm{M}})^2 + \boldsymbol{p}^2})$, therefore this θ-function can be omitted.

Next we present the off-shell result for the imaginary part of $\Sigma_4(\boldsymbol{p}, -\mathrm{i}p_0)$, performing again a partial-fraction decomposition of Eq. (5.13):

$$
\begin{aligned}
\Sigma_4(\boldsymbol{p}, -\mathrm{i}p_0) = {}& g_{\mathrm{Mqq}}^2\int\frac{\mathrm{d}^3q}{(2\pi)^3}\frac{1}{4p_0 E_b E_f}\left[\frac{(1 - n_{\mathrm{F}}^-(E_f) + n_{\mathrm{B}}(E_b))(E_b + p_0)}{E_f + E_b + p_0} +\right. \\
& + \frac{(n_{\mathrm{B}}(E_b) + n_{\mathrm{F}}^-(E_f))(p_0 - E_b)}{E_f - E_b + p_0 + \mathrm{i}\epsilon} + \frac{(n_{\mathrm{B}}(E_b) + n_{\mathrm{F}}^+(E_f))(p_0 - E_b)}{E_f - E_b - p_0 - \mathrm{i}\epsilon} + \left.\frac{(1 + n_{\mathrm{B}}(E_b) - n_{\mathrm{F}}^+(E_f))(p_0 - E_b)}{E_f + E_b - p_0 - \mathrm{i}\epsilon}\right].
\end{aligned} \tag{5.59}
$$

In comparison to Eq. (5.9) there is the factor p_0 in the denominator and also combinations of E_b and p_0 in the numerators, but Σ_4 features the very same pole structure as discussed before. Therefore we find immediately:

$$
\begin{aligned}
\mathrm{Im}\,\Sigma_4^{\mathrm{off}} = {}& \frac{g_{\mathrm{Mqq}}^2}{16\pi|\boldsymbol{p}|\,p_0}\left\{\int_{\mathrm{I}}\mathrm{d}E_f\,(-E_f)\left[-n_{\mathrm{B}}(E_f + p_0) - n_{\mathrm{F}}^-(E_f)\right] +\right. \\
& + \int_{\mathrm{II}}\mathrm{d}E_f\,E_f\left[n_{\mathrm{B}}(E_f - p_0) + n_{\mathrm{F}}^+(E_f)\right] + \\
& + \left.\int_{\mathrm{III}}\mathrm{d}E_f\,E_f\left[1 + n_{\mathrm{B}}(p_0 - E_f) - n_{\mathrm{F}}^+(E_f)\right]\right\} = \frac{g_{\mathrm{Mqq}}^2}{16\pi|\boldsymbol{p}|\,p_0}\left(K^{\mathrm{I}} + K^{\mathrm{II}} + K^{\mathrm{III}}\right).
\end{aligned} \tag{5.60}
$$

Introducing the two auxiliary functions

$$
\mathcal{G}^{\pm}(E) = T^2\left[\frac{\pi^2}{3} + \mathrm{Li}_2\left(1 - \mathrm{e}^{\beta(E \pm p_0)}\right) + \mathrm{Li}_2\left(-\mathrm{e}^{\beta(E \pm \mu)}\right)\right], \tag{5.61}
$$

we find:

$$
\begin{aligned}
K^{\mathrm{I}} = {}& \theta(|\boldsymbol{p}|-p_0)\left\{\frac{1}{2}(\mu^2 - p_0^2) - \widetilde{E}_- T\ln n_{\mathrm{F}}^-(\widetilde{E}_-) - p_0 T\ln n_{\mathrm{B}}(\widetilde{E}_- + p_0) + \mathcal{G}^+(\widetilde{E}_-)\right\} + \\
& + \theta(p_0 - |\boldsymbol{p}|)\theta(m_{\mathrm{M}} - m)\theta(\sqrt{(m-m_{\mathrm{M}})^2 + \boldsymbol{p}^2} - p_0)\left\{p_0 T\ln\frac{n_{\mathrm{B}}(\widetilde{E}_+ + p_0)}{n_{\mathrm{B}}(\widetilde{E}_- + p_0)} + \right. \\
& + \widetilde{E}_+ T\ln n_{\mathrm{F}}^-(\widetilde{E}_+) - \widetilde{E}_- T\ln n_{\mathrm{F}}^-(\widetilde{E}_-) + \left.\left(\mathcal{G}^+(\widetilde{E}_-) - \mathcal{G}^+(\widetilde{E}_+)\right)\right\},
\end{aligned} \tag{5.62}
$$

5.2. Quark self-energy from mesonic fluctuations

$$K^{\mathrm{II}} = \theta(|\boldsymbol{p}| - p_0) \left\{ \frac{1}{2}(\mu^2 - p_0^2) + \widetilde{E}_+ T \ln n_{\mathrm{F}}^+(-\widetilde{E}_+) + p_0 T \ln n_{\mathrm{B}}(-\widetilde{E}_+ - p_0) + \mathcal{G}^-(-\widetilde{E}_+) \right\} +$$
$$+ \theta(p_0 - |\boldsymbol{p}|)\theta(m - m_{\mathrm{M}})\theta(\sqrt{(m-m_{\mathrm{M}})^2 + \boldsymbol{p}^2} - p_0) \left\{ p_0 T \ln \frac{n_{\mathrm{B}}(-\widetilde{E}_+ - p_0)}{n_{\mathrm{B}}(-\widetilde{E}_- - p_0)} + \right.$$
$$\left. + \widetilde{E}_+ T \ln n_{\mathrm{F}}^+(-\widetilde{E}_+) - \widetilde{E}_- T \ln n_{\mathrm{F}}^+(-\widetilde{E}_-) + \left(\mathcal{G}^-(-\widetilde{E}_+) - \mathcal{G}^-(-\widetilde{E}_-) \right) \right\},$$
(5.63)

$$K^{\mathrm{III}} = \theta(p_0 - \sqrt{(m+m_{\mathrm{M}})^2 + \boldsymbol{p}^2}) \left\{ -p_0 T \ln \frac{n_{\mathrm{B}}(\widetilde{E}_+ + p_0)}{n_{\mathrm{B}}(\widetilde{E}_- + p_0)} + \right.$$
$$\left. + \widetilde{E}_- T \ln n_{\mathrm{F}}^-(\widetilde{E}_-) - \widetilde{E}_+ T \ln n_{\mathrm{F}}^-(\widetilde{E}_+) - \left(\mathcal{G}^+(\widetilde{E}_-) - \mathcal{G}^+(\widetilde{E}_+) \right) \right\}.$$
(5.64)

Note that the expression for K^{III} is just the negative of the second contribution to K^{I}.

Having derived results for Im Σ_0^{off} and Im Σ_4^{off}, the remaining integral for Σ_3 can be performed easily, since all building blocks have been prepared. The main observation is that $\mathcal{F}_3$ splits into two parts: the first one is independent of E_f, the second one introduces the same E_f dependence present in the calculation for Σ_4:

$$2\boldsymbol{p}^2 \mathcal{F}_3 = m_{\mathrm{M}}^2 + \boldsymbol{p}^2 + \boldsymbol{q}^2 - E_b^2 = \begin{cases} m_{\mathrm{M}}^2 + \boldsymbol{p}^2 - p_0^2 - m^2 - 2E_f p_0 & \text{for Case I}, \\ m_{\mathrm{M}}^2 + \boldsymbol{p}^2 - p_0^2 - m^2 + 2E_f p_0 & \text{for Case II and Case III}. \end{cases}$$
(5.65)

We find therefore

$$\mathrm{Im}\, \Sigma_3^{\mathrm{off}} = \frac{g_{\mathrm{Mqq}}^2}{16\pi |\boldsymbol{p}|} \left\{ \int_{\mathrm{I}} \mathrm{d}E_f\, \mathcal{F}_3^{\mathrm{I}} \left[-n_{\mathrm{B}}(E_f + p_0) - n_{\mathrm{F}}^-(E_f) \right] + \right.$$
$$+ \int_{\mathrm{II}} \mathrm{d}E_f\, \mathcal{F}_3^{\mathrm{II,III}} \left[n_{\mathrm{B}}(E_f - p_0) + n_{\mathrm{F}}^+(E_f) \right] +$$
$$\left. + \int_{\mathrm{III}} \mathrm{d}E_f\, \mathcal{F}_3^{\mathrm{II,III}} \left[1 + n_{\mathrm{B}}(p_0 - E_f) - n_{\mathrm{F}}^+(E_f) \right] \right\}.$$
(5.66)

Inspecting the definitions for J^x in Eq. (5.55) and K^x in Eq. (5.60), we find the relation:

$$\mathrm{Im}\, \Sigma_3^{\mathrm{off}} = \frac{m_{\mathrm{M}}^2 + \boldsymbol{p}^2 - p_0^2 - m^2}{2\boldsymbol{p}^2} \mathrm{Im}\, \Sigma_0^{\mathrm{off}} + \frac{p_0^2}{\boldsymbol{p}^2} \mathrm{Im}\, \Sigma_4^{\mathrm{off}},$$
(5.67)

where we recall

$$\mathrm{Im}\, \Sigma_0^{\mathrm{off}} = \frac{g_{\mathrm{Mqq}}^2}{16\pi |\boldsymbol{p}|} \left(J^{\mathrm{I}} + J^{\mathrm{II}} + J^{\mathrm{III}} \right),$$
$$\mathrm{Im}\, \Sigma_4^{\mathrm{off}} = \frac{g_{\mathrm{Mqq}}^2}{16\pi |\boldsymbol{p}| p_0} \left(K^{\mathrm{I}} + K^{\mathrm{II}} + K^{\mathrm{III}} \right).$$
(5.68)

The resulting off-shell shear viscosity will be discussed in Section 6.3. It is interesting to note that in contrast to the on-shell results in Eqs. (5.38) and (5.40), the off-shell imaginary parts feature a vacuum contribution generated by Case III. For $T, \mu \to 0$, all distribution functions vanish but the 1-term in the integrand contributes, cf. Eqs. (5.55), (5.60), and (5.66).

Apart from the vacuum limit it is instructive to check the on-shell limit which is included in the off-shell results. The on-shell case is taken when setting $s = p_0^2 - \boldsymbol{p}^2 = m^2 > 0$. Case II and Case III cannot contribute and only Case I has to be investigated. As we have already discussed,

5. Mesonic fluctuations in the quark sector

the boundaries $\widetilde{E}_\pm$ convert into their corresponding on-shell expressions, see Eq. (5.46). We find:

$$\mathrm{Im}\,\Sigma_0^{\mathrm{off}}\bigg|_{\mathrm{on-shell}} = \frac{g_{\mathrm{Mqq}}^2}{16\pi|\boldsymbol{p}|}J^{\mathrm{I}}\bigg|_{\mathrm{on-shell}} = \frac{g_{\mathrm{Mqq}}^2}{16\pi|\boldsymbol{p}|}T\ln\frac{n_{\mathrm{F}}^-(E_{\mathrm{min}})\,n_{\mathrm{B}}(E_{\mathrm{max}}+p_0)}{n_{\mathrm{F}}^-(E_{\mathrm{max}})\,n_{\mathrm{B}}(E_{\mathrm{min}}+p_0)}, \quad (5.69)$$

which is just $\mathrm{Im}\,\Sigma_0$ given in Eq. (5.38). We have omitted the θ-functions, but note that the usual on-shell condition $m_{\mathrm{M}} > 2m$ follows automatically for $s = p_0^2 - \boldsymbol{p}^2 = m^2$:

$$\begin{aligned}
\theta(m_{\mathrm{M}}-m)\theta(p_0-|\boldsymbol{p}|)\theta(\sqrt{(m_{\mathrm{M}}-m)^2}-p_0)\Big|_{\mathrm{on-shell}} &= \\
= \theta(m_{\mathrm{M}}-m)\theta(m_{\mathrm{M}}^2 - 2m_{\mathrm{M}}m + m^2 + \boldsymbol{p}^2 - (\boldsymbol{p}^2+m^2)) &= \\
= \theta(m_{\mathrm{M}}-m)\theta(m_{\mathrm{M}}(m_{\mathrm{M}}-2m)) &= \\
= \theta(m_{\mathrm{M}} - 2m)\,.
\end{aligned} \quad (5.70)$$

In order to check the on-shell limits for $\mathrm{Im}\,\Sigma_{3,4}$, we first note the identity

$$\mathcal{H}(E) = (E+p_0)\ln n_{\mathrm{F}}^-(E) - \frac{1}{T}\left(\mathcal{G}^+(E) - \frac{\pi^2 T^2}{3}\right), \quad (5.71)$$

where $\mathcal{H}(E)$ has been defined in Eq. (5.39) and $\mathcal{G}^+$ in Eq. (5.61). Note that the constant $\pi^2 T^2/3$ is not relevant in the on-shell results, since there only the differences $\mathcal{H}(E_{\mathrm{max}}) - \mathcal{H}(E_{\mathrm{min}})$ occurs. We find:

$$\begin{aligned}
K^{\mathrm{I}}\Big|_{\mathrm{on-shell}} &= p_0 T \ln\frac{n_{\mathrm{F}}^-(E_{\mathrm{min}})\,n_{\mathrm{B}}(E_{\mathrm{max}}+p_0)}{n_{\mathrm{F}}^-(E_{\mathrm{max}})\,n_{\mathrm{B}}(E_{\mathrm{min}}+p_0)} + \\
&+ (E_{\mathrm{max}}+p_0)T\ln n_{\mathrm{F}}^-(E_{\mathrm{max}}) - (E_{\mathrm{min}}+p_0)T\ln n_{\mathrm{F}}^-(E_{\mathrm{min}}) + \mathcal{G}^+(E_{\mathrm{min}}) - \mathcal{G}^+(E_{\mathrm{max}}) \\
&= \frac{16\pi|\boldsymbol{p}|p_0}{g_{\mathrm{Mqq}}^2}\,\mathrm{Im}\,\Sigma_0 + T\left(\mathcal{H}(E_{\mathrm{max}}) - \mathcal{H}(E_{\mathrm{min}})\right),
\end{aligned} \quad (5.72)$$

which is in agreement with the on-shell result for $\mathrm{Im}\,\Sigma_4$ given in Eq. (5.40).

5.3. Vacuum fluctuations and the cloudy bag model

In order to investigate the vacuum limit of the thermal self-energy in Eq. (5.7), we independently calculate the quark self-energy in Minkowski space, i.e. at $T = 0$ and $\mu = 0$. With the incoming quark four-momentum, $p = (p_0, \boldsymbol{p})$, we find

$$\begin{aligned}
\Sigma^{\mathrm{S/P}}(p^2) &= ig_{\mathrm{Mqq}}^2 \int \frac{\mathrm{d}^4 q}{(2\pi)^4}\,\Gamma^{\mathrm{S/P}}\frac{1}{\slashed{q}-m}\Gamma^{\mathrm{S/P}}\frac{1}{(q-p)^2 - m_{\mathrm{M}}^2} = \\
&= \mp m\,\Sigma_0^{\mathrm{vac}}(p^2) + \slashed{p}\,\Sigma_1^{\mathrm{vac}}(p^2) = \\
&= \mp m\,\Sigma_0^{\mathrm{vac}} + \boldsymbol{p}\cdot\boldsymbol{\gamma}\,\Sigma_3^{\mathrm{vac}} - p_0\gamma_0\,\Sigma_4^{\mathrm{vac}}\,.
\end{aligned} \quad (5.73)$$

We have decomposed the Σ_1^{vac} contribution into momentum and energy parts, Σ_3^{vac} and Σ_4^{vac}, respectively, since the thermal medium breaks Lorentz invariance and introduces three instead of two contributions to the quark self energy. However, in the vacuum we expect $\Sigma_3^{\mathrm{vac}}(p^2) = \Sigma_4^{\mathrm{vac}}(p^2)$, though these two terms are divergent. We find:

$$\Sigma_0^{\mathrm{vac}}(p^2) = -ig_{\mathrm{Mqq}}^2 \int \frac{\mathrm{d}^4 q}{(2\pi)^4}\frac{1}{q_0^2 - E_f^2}\frac{1}{(q_0-p_0)^2 - E_b^2} = g_{\mathrm{Mqq}}^2 \int \frac{\mathrm{d}^3 q}{(2\pi)^3}\frac{E_+}{2E_b E_f(E_+^2 - p_0^2)}, \quad (5.74)$$

5.3. Vacuum fluctuations and the cloudy bag model

where the convergent q_0 integration is performed using residue calculus. The remaining three-dimensional integral suffers from divergency and is understood to be regularized by the NJL cutoff Λ. We emphasize that in the vacuum such a Lorentz-symmetry breaking regularization scheme is usually not applied. Instead, dimensional regularization is used preserving the (classical) symmetries of the Lagrangian. To compare with the vacuum limit of the thermal results, the Matsubara sum is replaced by carrying out the q_0 integration. The Lorentz covariance is ensured for large cutoff values, $\Lambda \to \infty$, therefore it is justified to write Σ_0^{vac} as function of $p^2 = p \cdot p$ only. We also find:

$$\Sigma_3^{\text{vac}}(p^2) = -ig_{\text{Mqq}}^2 \int \frac{d^4q}{(2\pi)^4} \mathcal{F}_3 \frac{1}{p_0^2 - E_f^2} \frac{1}{(q_0 - p_0)^2 - E_b^2} = g_{\text{Mqq}}^2 \int \frac{d^3q}{(2\pi)^3} \frac{\mathcal{F}_3 E_+}{2E_b E_f (E_+^2 - p_0^2)} ,$$

$$\Sigma_4^{\text{vac}}(p^2) = -ig_{\text{Mqq}}^2 \int \frac{d^4q}{(2\pi)^4} \frac{q_0}{p_0} \frac{1}{p_0^2 - E_f^2} \frac{1}{(q_0 - p_0)^2 - E_b^2} = g_{\text{Mqq}}^2 \int \frac{d^3q}{(2\pi)^3} \frac{1}{2E_b(E_+^2 - p_0^2)} .$$
(5.75)

Despite the fact that again both self-energy contributions are Lorentz covariant for large cutoff values, they are different: $\Sigma_4^{\text{vac}} > \Sigma_3^{\text{vac}}$. Usually, one expects that this difference converges to zero when the cutoff reaches large values. In contrast, we find also in this limit some finite difference between the two Dirac parts:

$$\lim_{\Lambda \to \infty} \frac{8\pi^3}{g_{\text{Mqq}}^2} (\Sigma_4^{\text{vac}} - \Sigma_3^{\text{vac}}) = \lim_{\Lambda \to \infty} \int d^3q \frac{E_f - \mathcal{F}_3 E_+}{2E_b E_f (E_+^2 - p_0^2)} = \frac{\pi}{6} \neq 0 . \quad (5.76)$$

The fact that the difference between the two Dirac terms does not converge to zero is a shortcoming of the regularization scheme we applied. Using the three-momentum cutoff instead of a symmetry-conserving regularization scheme induces some remaining pieces also for $\Lambda \to \infty$. However, since in the Matsubara formalism the Matsubara sum needs to be carried out, the three-momentum cutoff is the only suitable procedure to compare Minkowski-space results to thermal results in their vacuum limit. Indeed, the limits match the calculations in Minkowski space:

$$\lim_{T,\mu \to 0} \Sigma_i(\boldsymbol{p}, \nu_n) \Big|_{\nu_n \mapsto -ip_0} = \Sigma_i^{\text{vac}}(p_0, \boldsymbol{p}) , \quad \text{for } i = 0, 3, 4 . \quad (5.77)$$

With the sign conventions from Eqs. (5.7) and (5.73) one finds:

$$\lim_{T,\mu \to 0} \Sigma_\beta^{\text{S/P}}(\boldsymbol{p}, \nu_n) \Big|_{\nu_n \mapsto -ip_0} = -\Sigma^{\text{S/P}}(p^2) . \quad (5.78)$$

The vacuum mass shift is calculated by standard means[42], e.g. [PS95]:

$$\delta m^{\text{S/P,vac}} = \Sigma \big|_{\not{p} \to m} = \mp m \Sigma_0^{\text{vac}}(m^2) - m \Sigma_4^{\text{vac}}(m^2) = g_{\text{Mqq}}^2 m \big(\mp J_0(m) - J_4(m) \big) . \quad (5.79)$$

For convenience, the two integrals $J_i(E) = g_{\text{Mqq}}^{-2} \Sigma_i(E^2)$ for $i = 0, 4$ have been introduced as

$$J_0(E) = \int \frac{d^3q}{(2\pi)^3} \frac{1}{4E_b E_f} \left[\frac{1}{E_b + E_f - E} + \frac{1}{E_b + E_f + E} \right] > 0 ,$$

$$J_4(E) = \int \frac{d^3q}{(2\pi)^3} \frac{1}{4E_b E} \left[\frac{1}{E_b + E_f - E} - \frac{1}{E_b + E_f + E} \right] > 0 ,$$
(5.80)

---

[42]Note that the vacuum mass shift also follows from our more general analysis in the last section when interpreting $\not{p} \mapsto m$ as setting $\boldsymbol{p} \mapsto 0$ in Ω in Eq. (5.16) and evaluating $\delta m^{\text{S/P}}$ as defined in Eq. (5.17).

5. Mesonic fluctuations in the quark sector

from which one gets

$$(J_0 - J_4)(E) = \int \frac{d^3q}{(2\pi)^3} \frac{1}{4E_f E} \left[\frac{1}{E_b + E_f - E} - \frac{1}{E_b + E_f + E} \right] > 0 . \tag{5.81}$$

In conclusion, the mesonic fluctuations introduce corrections to the vacuum constituent-quark mass by screening (pion case) and antiscreening effects (sigma-meson case):

$$\begin{aligned} 0 < \delta m^{\pi,\text{vac}} &= 3g_{\pi qq}^2 m(J_0(m) - J_4(m)) = 60.1 \text{ MeV} , \\ 0 > \delta m^{\sigma,\text{vac}} &= -g_{\sigma qq}^2 m(J_0(m) + J_4(m)) = -21.1 \text{ MeV} . \end{aligned} \tag{5.82}$$

In total, the mesonic corrections to the constituent-quark mass $m^{\text{vac}} = 325$ MeV is only 12%. This rather small correction is consistent with the treatment of the mesonic fluctuations as a $1/N_c$-suppressed Fock term in the gap equation (5.1). Our results compare well to those from [QK94], where this correction has been determined to be 16%, but using a slightly different NJL parameter set and restricting the entire discussion to the vacuum case only. The qualitative screening and antiscreening effects have been found as well.

Our full field-theoretical results for the vacuum mass shifts also allow to rederive well-known results from the cloudy bag model, cf. for instance [HK78, TTM81, Tho84, HT96]. Corrections to the hadron masses (in particular to the nucleon mass) are derived from a model where quarks are moving freely inside the bag with radius R. The reflection of the free quarks on the bag surface provides some naive model of confinement. However, the boundary condition of this MIT bag model violates chiral symmetry since the helicity of a (massless) quark changes from $+1$ to -1 when being reflected at the surface. One can overcome this issue and extend the MIT bag model by including a pion cloud which surrounds the confined quark core. Only at the bag surface interaction between the quarks and the pions can happen. This leads to the (chiral) cloudy bag model which Lagrangian is given by [HT96, TW01]

$$\mathcal{L}_{\text{CBM}} = \underbrace{\left(i\bar{\psi}\gamma_\mu \partial^\mu \psi - m\bar{\psi}\psi - B\right) \Theta(R - r) - \frac{1}{2}\delta_S \bar{\psi}\psi}_{\text{free quark with reflection term}} + \underbrace{\frac{i\delta_S}{4f_\pi} \bar{\psi}\gamma_5 \boldsymbol{\tau}\psi \cdot \boldsymbol{\pi}}_{\text{interaction term}} + \underbrace{\frac{1}{2}(\partial_\mu \boldsymbol{\pi})^2 - \frac{1}{2}m_\pi^2 \boldsymbol{\pi}^2}_{\text{free pion}} . \tag{5.83}$$

One has introduced δ_S being a delta-function peaking at the bag surface and the bag (energy) constant B. Our aim is to compare the MIT result for the mass shift of the nucleon to our result for the mass shift for the constituent-quark induced by the pion cloud, $\delta m^{\pi,\text{vac}}$. In the MIT bag model one finds for the nucleon mass at second-order perturbation theory [TW01]

$$M_N = M_N^{(0)} + \delta M_N , \tag{5.84}$$

with

$$\delta M_N = -\frac{3g_{\pi qq}^2}{16\pi^2 M_N^2} \int_0^\infty dq \frac{q^4 u_{NN}^2(q)}{\omega^2(q)} < 0 , \tag{5.85}$$

where $\omega(q)$ denotes the nucleon energy and $u_{NN}(q)$ is a momentum-dependent function induced by the spherical geometry present in the (MIT) bag model. As always, in second-order perturbation theory, the mass correction $\delta M_N < 0$ is negative, therefore the bare nucleon mass, $M_N^{(0)}$, is slightly larger than the one including the pion cloud. Within the NJL model, in contrast, we have found a screening effect of the pion cloud, $\delta m^{\pi,\text{vac}} > 0$, i.e. the (Hartree) constituent-quark mass is slightly smaller than the one including the pion cloud. This qualitative difference between the NJL and MIT bag model can be explained by the fact that second-order perturbation theory does not include the full field-theoretical interaction between quarks and pseudoscalar

5.3. Vacuum fluctuations and the cloudy bag model

mesons, in particular, purely relativistic effects are not included in the MIT bag model. This can be seen by expanding our results in its non-relativistic limit, i.e. expanding $\delta m^{\pi,\text{vac}}$ in $1/m$:

$$\delta m^{\pi,\text{vac}} = \frac{g_{\pi qq}^2}{8\pi^2 m^2} \int dq\, q^2 \left(\frac{m}{E_b} + \frac{1}{2} - \frac{q^2}{2E_b^2} + \ldots \right). \tag{5.86}$$

The first two terms in the integrand contribute with a positive sign to the screening effect found in the NJL model. The third term, which is not dominant in the limit $m \to \infty$, weakens the screening effect by some negative contribution. This term coincides with the nucleon mass shift from the MIT bag model:

$$-\frac{g_{Mqq}^2}{8\pi^2 m^2} \int dq\, \frac{q^4}{2E_b^2} = \delta M_N \big|_{u_{NN}(k)\mapsto 1,\, \omega(k)\mapsto E_b,\, M_N \mapsto m}. \tag{5.87}$$

In conclusion, the NJL results for the mass shift of the constituent quark includes in its non-relativistic expansion the results from the MIT bag model. They are qualitatively different in their sign of the mass shift, and the overall screening effect in the NJL model is explained by the overcompensation of the non-relativistic effects by the leading-orders in an expansion in the inverse constituent-quark mass.

6. The ratio η/s in the NJL model

> "Who would have thought around 1900 that in fifty years time we would know so much more and understand so much less." [Lan65]
>
> Albert Einstein

In this chapter we present our main results for the temperature dependence of the ratio η/s. We first discuss how the entropy density s can be derived within the NJL model applying again a large-N_c analysis. Results for η/s as function of temperature and quark chemical potential are shown. The full thermal dependencies of all parameters such as constituent-quark and meson masses but also the quark-meson coupling constant are incorporated. The on-shell results are restricted to the temperature range above the Mott temperature, $T > T_M$. This shortcoming will be removed when relaxing the on-shell condition and allowing also for off-shell contributions to the imaginary parts of the Dirac self energies.

6.1. Entropy density in a large-N_c expansion

All thermodynamic quantities can be derived from the NJL partition function Z which is decomposed into free and interaction parts,

$$\ln Z = \ln Z_0 + \ln Z_{\text{int}} = \ln Z_0 + \sum_{k=1}^{\infty} \ln Z_k \,, \tag{6.1}$$

where the interaction part is again treated within a large-N_c expansion. In Section 3.3.2 we have investigated the (2PI) generating functional Φ which determines also the partition function:

$$\ln Z_k = \beta V \, \Phi^{(k-1)} \,. \tag{6.2}$$

The prefactor V denotes the three-dimensional volume which drops out for intensive quantities like entropy density s or pressure P. Note that $\ln Z_k \sim N_c^{2-k}$ is only true for $k \leq 1$, because for the non-interacting case $\ln Z_0$, there is no Feynman-diagram representation [Hel11]. It turns out that the first two terms, $\ln Z_0$ and $\ln Z_1$, scale linearly with N_c.

We first consider the non-interacting part of the partition function. In the limit $N_c \to \infty$ the NJL model simply becomes a free Fermi theory because the four-fermion coupling becomes small, $G \sim 1/N_c$, and one has [KG06]:

$$\ln Z_0 = \frac{N_c N_f V}{\pi^2} \int_0^{\Lambda,\infty} dp\, p^2 \left[\beta E + \ln\left(1 + e^{-\beta(E-\mu)}\right) + \ln\left(1 + e^{-\beta(E+\mu)}\right) \right], \tag{6.3}$$

with E being the quark energy. As we have mentioned, one finds the scaling $\ln Z_0 \sim N_c$. Note that for ensuring the Stefan Boltzmann limit of thermodynamic quantities at high temperatures one has to apply the *soft-cutoff scheme* [Bra13], denoted by $\int^{\Lambda,\infty} dp$: the momentum integral contains the thermal constituent-quark mass $m(T)$ for $p \in [0, \Lambda]$ but for $p > \Lambda$ the NJL coupling G drops to zero and the quarks feature the current-quark mass m_0 only.

6. The ratio η/s in the NJL model

Thermodynamic quantities are deduced from the partition function by standard means, e.g. the pressure or energy density are derived as

$$P = \frac{\ln Z}{\beta V}, \quad \text{and} \quad s = \frac{\partial S}{\partial V} = -\frac{\partial^2 F}{\partial V \partial T} = \frac{\partial P}{\partial T}, \tag{6.4}$$

with F denoting the free energy (4.53). In the massless limit (Stefan Boltzmann limit) one has $E = p$ and the momentum integral for $\ln Z_0$ can be carried out resulting in:

$$s_{\text{SB}} = \lim_{m_0 \to 0} s_0 = -\frac{2N_c N_f T^2}{\pi^2} \left[\mu \left(\text{Li}_3(-e^{-\beta \mu}) - \text{Li}_3(-e^{\beta \mu}) \right) + 4T \left(\text{Li}_4(-e^{-\beta \mu}) + \text{Li}_4(-e^{\beta \mu}) \right) \right] =$$
$$= \frac{1}{3} N_c N_f \left(\frac{7\pi^2}{15} T^3 + T\mu^2 \right). \tag{6.5}$$

Note that the divergence of $\ln Z_0$ coming from the first term in the integrand of Eq. (6.3), $d^3p\,\beta E$, contributes only to the pressure but does not affect the entropy density. This comes from the fact that the temperature dependence is canceled before taking the derivative with respect to T. In the first line of Eq. (6.5) we have used the polylogarithm (polylog) function defined by

$$\text{Li}_n(z) = \sum_{k=1}^{\infty} \frac{z^k}{k^n}. \tag{6.6}$$

It is connected to Riemann's zeta function via $\text{Li}_n(1) = \zeta(n)$ and $\text{Li}_n(-1) = \left(2^{1-n} - 1 \right) \zeta(n)$. We emphasize the interesting property that the rather complicated combinations of polylog functions sum up to a simple polynomial expression in μ and T:

$$\begin{aligned}
\text{Li}_3(-e^{-x}) - \text{Li}_3(-e^{x}) &= \frac{\pi^2}{6} + \frac{1}{6} x^3, \\
\text{Li}_4(-e^{-x}) + \text{Li}_4(-e^{x}) &= -\frac{7\pi^4}{360} - \frac{\pi^2}{12} x^2 - \frac{1}{24} x^4.
\end{aligned} \tag{6.7}$$

Results for the entropy density, s, compared to the Stefan-Boltzmann limit are shown in Fig. 6.1 where we have used again the thermal constituent-quark mass m. In the low-temperature region the constituent-quark mass is large, $m \gg T$, leading to a suppression of s until $T \lesssim 200$ MeV. With increasing temperature the chiral condensate $\langle \bar\psi \psi \rangle$ is melting, therefore one approaches the Stefan-Boltzmann limit for $T \to \infty$. If one uses the current-quark mass instead the resulting entropy density is very close to the Stefan-Boltzmann limit and only for very small temperatures, $T \lesssim m_0$, deviations from this massless limit are visible.

Taking also interactions of the constituent quarks into account, i.e. calculating the leading-order term $\ln Z_1$ of the interaction part, one finds with:

$$\ln Z_1 = \frac{G\beta V}{2} \left[T \sum_{n \in \mathbb{Z}} \int \frac{d^3 p}{(2\pi)^3} \text{Tr}\, G^{\text{F}}_{\beta}(\boldsymbol{p}, \nu_n) \right]^2 =$$
$$= \frac{2G\beta V}{\pi^4} N_c^2 N_f^2 \left[\int_0^{\Lambda} dp\, \frac{p^2 m^2}{E} \left(1 - n_{\text{F}}^+(E) - n_{\text{F}}^-(E) \right) \right]^2, \tag{6.8}$$

where we have carried out the Matsubara sum using the master formula (A.21) listed in the Appendix. As stated above, this term scales as $\ln Z_1 \sim GN_c^2 \sim N_c$, which is the same linear dependence as the non-interacting partition function. All higher order corrections to the partition function are suppressed in a large-N_c expansion. We also mention that the soft cutoff scheme introduced in Section 3.4 reduces for $\ln Z_1$ to the simple NJL cutoff because there is the

6.1. Entropy density in a large-N_c expansion

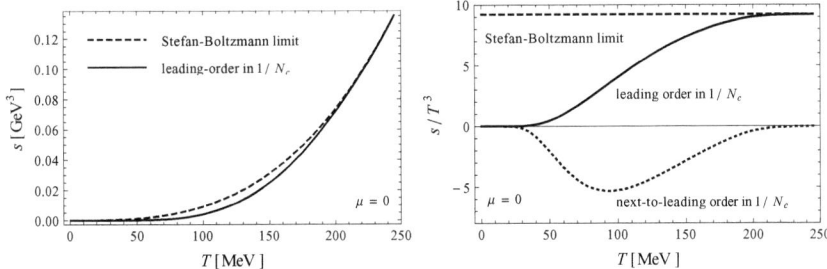

Figure 6.1.: (Reduced) entropy density at leading order (left) and next-to-leading order (right) in comparison to the Stefan-Boltzmann limit of a free fermion gas (dashed lines)

coupling G in the prefactor and one has $G(p > \Lambda) = 0$. Since one has

$$\frac{\partial}{\partial T}\left(1 - n_F^+(E) - n_F^-(E)\right) < 0 , \tag{6.9}$$

the next-to-leading order correction to the entropy density turns out to be negative: $s_1(T,\mu) < 0$. In combination with the non-interaction part derived from $\ln Z_0$ the total entropy density remains positive: $s = s_0 - |s_1| > 0$. We show our next-to-leading NJL result for the entropy density in Fig. 6.1. Having a brief look to other field theories and models, we realize that it is a common pattern that (attractive) interactions lead to a negative next-to-leading-order correction to the entropy density [KG06]:

$$s_{\lambda\phi^4}(T) = \frac{2\pi^2 T^3}{45}\left[1 - \frac{15\lambda}{8\pi^2}\right] + \ldots$$

$$s_{\text{QED}}(T) = \frac{11\pi^2 T^3}{45}\left[1 - \frac{25\alpha}{22\pi}\right] + \ldots \tag{6.10}$$

$$s_{\text{QCD}}(T) = 4d_A T^3\left[\frac{1}{5}\left(1 + \frac{7 d_F}{4 d_A}\right) - \frac{\alpha_s^2}{4\pi}\left(C_A + \frac{5}{2}S_F\right)\right] + \ldots$$

Also in chiral perturbation theory this pattern can be observed when expanding the entropy density in inverse powers of f_π [GL89]:

$$s_{\chi\text{PT}}(T) = \frac{T}{2\pi^2}\left[4T^2 h_5(\beta m_\pi) + 3m_\pi^2 h_3(\beta m_\pi)\right] - \frac{3m_\pi^2 T}{16\pi^4 f_\pi^2} h_3(\beta m_\pi)\left[2T^2 h_3(\beta m_\pi) + m_\pi^2 h_1(\beta m_\pi)\right], \tag{6.11}$$

where one introduces the positive function

$$h_n(\xi) = \int_\xi^\infty dx \, \frac{(x^2 - \xi^2)^{\frac{n}{2}-1}}{e^x - 1} . \tag{6.12}$$

However, for our purposes the resulting entropy density we use for the numerical evaluation of the ratio η/s in the next sections finally calculates to

$$s(T,\mu) = \frac{N_c N_f}{\pi^2}\int_0^{\Lambda,\infty} dp\, p^2 \left[-\ln n_F^+(E) - \ln n_F^-(E) + \beta(E+\mu)n_F^+(E) + \beta(E-\mu)n_F^-(E)\right]. \tag{6.13}$$

6.2. Kubo formalism for the Dirac self-energy

In Section 4.2 we have discussed the shear viscosity $\eta[\Gamma(p)]$ assuming the parameterization of the full quark propagator given in Eq. (4.32). As we have demonstrated in Chapter 5, the $1/N_c$ corrections from mesonic fluctuations to the quark propagator give rise to a richer Dirac structure as assumed in the quasiparticle ansatz, cf. Eqs. (5.7) and (5.34). Three imaginary parts instead of just one enter the Kubo formula given in Eq. (4.30) in terms of the spectral function $\rho(\epsilon, \boldsymbol{p})$. Its Dirac structure can be parameterized by three functions A, B, C, cf. Appendix A.3, where we denote their denominator by D:

$$\rho(\epsilon, \boldsymbol{p}) = -\frac{1}{\pi D} \left[mA + p_0 \gamma_0 B - \boldsymbol{p} \cdot \boldsymbol{\gamma} C \right]. \tag{6.14}$$

These four functions depend on the (off-shell) energy ϵ, the three-momentum $\boldsymbol{p}$, and the thermal parameters T and μ. They can be determined from the full quark propagator calculated within the NJL model in Chapter 5:

$$G_R(p_0, \boldsymbol{p}) = \frac{1}{\slashed{p} - m - \Sigma^{\text{tot}}} = \frac{m(1+\Sigma_0^{\text{tot}}) + p_0\gamma_0(1+\Sigma_4^{\text{tot}}) - \boldsymbol{p}\cdot\boldsymbol{\gamma}(1+\Sigma_3^{\text{tot}})}{p_0^2(1+\Sigma_4^{\text{tot}})^2 - \boldsymbol{p}^2(1+\Sigma_3^{\text{tot}})^2 - m^2(1+\Sigma_0^{\text{tot}})^2}, \tag{6.15}$$

where all mesonic contributions from different modes have been summed up taken their multiplicities into account:

$$\Sigma_j^{\text{tot}} = 3\Sigma_j^\pi + \Sigma_j^\sigma, \quad \text{for} \quad j = 0, 3, 4. \tag{6.16}$$

The quark self-energies from pionic and sigma fluctuations, $\Sigma_j^{\pi,\sigma}$, have been defined in Eq. (5.7) and their imaginary parts are given in Eq. (5.34). For the following calculation we now take only the relevant imaginary parts into account, i.e. we define

$$\operatorname{Im} \Sigma_j^{\text{tot}} = \rho_j, \quad \text{and} \quad \operatorname{Re} \Sigma_j^{\text{tot}} \equiv 0. \tag{6.17}$$

Not taking the real parts of the self-energy contributions into account is presumably a rather rough approximation violating the Kramers-Kronig relations for Σ_j^{tot}. Doing so, we ignore the momentum-dependence of the constituent-quark mass which appears at Fock level only. We therefore keep the constituent-quark masses at Hartree level. There, the imaginary part would simply vanish, cf. the gap equation (5.1). Formally, this approximation is equivalent to readjusting the NJL parameter set and introducing a new set $(m_0, G, \Lambda)^{\text{new}}$ that will depend on the thermal variables T and μ, and on energy and momentum, p_0 and $\boldsymbol{p}$, respectively. In conclusion, we use this approximation for simplicity and find for the full quark propagator:

$$G_R(p_0, \boldsymbol{p}) \approx \frac{(m + p_0\gamma_0 - \boldsymbol{p}\cdot\boldsymbol{\gamma}) + i(m\rho_0 + p_0\gamma_0\rho_4 - \boldsymbol{p}\cdot\boldsymbol{\gamma}\rho_3)}{[p_0^2(1-\rho_4^2) - \boldsymbol{p}^2(1-\rho_3^2) - m^2(1-\rho_0^2)] + 2i(p_0^2\rho_4 - \boldsymbol{p}^2\rho_3 - m^2\rho_0)}. \tag{6.18}$$

Introducing two auxiliary functions,

$$\begin{aligned} N_1 &= p_0^2(1-\rho_4^2) - \boldsymbol{p}^2(1-\rho_3^2) - m^2(1-\rho_0^2), \\ N_2 &= p_0^2\rho_4 - \boldsymbol{p}^2\rho_3 - m^2\rho_0, \end{aligned} \tag{6.19}$$

we identify the four functions parameterizing the quark spectral function $\rho = -\frac{1}{\pi}\operatorname{Im} G_R$ defined in Eq. (6.14):

$$A = \rho_0 N_1 - 2N_2, \quad B = \rho_4 N_1 - 2N_2, \quad C = \rho_3 N_1 - 2N_2, \quad \text{and} \quad D = N_1^2 + 4N_2^2. \tag{6.20}$$

6.2. Kubo formalism for the Dirac self-energy

The evaluation of the shear viscosity is now reduced to carrying out the Dirac trace present in the integrand of Eq. (4.30):

$$\begin{aligned}\text{Tr}[\gamma_2 \rho \gamma_2 \rho] &= \frac{1}{\pi^2 D^2} \left[\gamma_2 (mA + p_0 \gamma_0 B - \boldsymbol{p} \cdot \boldsymbol{\gamma} C) \gamma_2 (mA + p_0 \gamma_0 B - \boldsymbol{p} \cdot \boldsymbol{\gamma} C) \right] = \\ &= \frac{4}{\pi^2 D^2} \left[-m^2 A^2 + p_0^2 B^2 - \boldsymbol{p}^2 C^2 + 2 p_y^2 C^2 \right].\end{aligned} \quad (6.21)$$

The shear viscosity therefore reads:

$$\eta = \frac{2 N_c N_f}{3 \pi^2 T} \int_{-\infty}^{\infty} d\epsilon \int d^3 p \, n_F^+(\epsilon)(1 - n_F^+(\epsilon)) \left[\frac{p^4}{D^2} \cdot \frac{2}{5} \boldsymbol{p}^2 C^2 + \frac{p^4}{D} \left(-m^2 \rho_0^2 - \boldsymbol{p}^2 \rho_3^2 + p_0^2 \rho_4^2 \right) \right]. \quad (6.22)$$

As we have discussed in Section 4.2, this kind of integral features a sharp peak structure appearing when $D \approx 0$. Therefore, we have approximated this structure in Eq. (4.35) by taking only the dominant term $\sim D^{-2}$ into account and dropping the term $\sim D_{-1}$. Doing so, only the functions C and D are relevant and we find our previous result for the shear viscosity assuming the parameterization (4.32). We find indeed:

$$\begin{aligned}\lim_{\rho_3 \to 0} \lim_{\rho_4 \to 0} C^2(\epsilon, \boldsymbol{p}) \bigg|_{p_0 \mapsto -\Gamma/m} &= 4 m^2 \Gamma^2, \\ \lim_{\rho_3 \to 0} \lim_{\rho_4 \to 0} D(\epsilon, \boldsymbol{p}) \bigg|_{p_0 \mapsto -\Gamma/m} &= X(p)|_{p^2 = \epsilon^2 - \boldsymbol{p}^2},\end{aligned} \quad (6.23)$$

with $X(p)$ has been defined in Eq. (4.34). From this we conclude:

$$\lim_{\rho_3 \to 0} \lim_{\rho_4 \to 0} \eta \bigg|_{p_0 \mapsto -\Gamma/m} = (4.37). \quad (6.24)$$

One rediscovers the previous results for the shear viscosity when setting some Dirac parts of the quark self energy, ρ_3 and ρ_4, to zero, cf. Eq. (5.7). In this case one simply has

$$\Sigma_\beta^{\text{S/P}} = \pm \Gamma, \quad (6.25)$$

with no additional Dirac structure.

The more general result (6.22) includes in some limit the shear viscosity discussed in Section 6.22. The Kubo formula we use for evaluating the shear viscosity incorporates all three contributions from the Dirac part. We also avoid the peak-structure approximation and take both the D^{-2} and D^{-1} terms into account[43]. Our final result for the shear viscosity reads:

$$\eta = \frac{2 N_c N_f}{3 \pi^3 T} \int_{-\infty}^{\infty} d\epsilon \int dp \, n_F^+(\epsilon)(1 - n_F^+(\epsilon)) \frac{p^4}{D^2(\epsilon, \boldsymbol{p})} \left[-m^2 A^2(\epsilon, \boldsymbol{p}) - \frac{3}{5} \boldsymbol{p}^2 C^2(\epsilon, \boldsymbol{p}) + p_0^2 B^2(\epsilon, \boldsymbol{p}) \right]. \quad (6.26)$$

It is remarkable how the negative and positive terms in the integrand balance to yield the overall positive shear viscosity $\eta > 0$. The functions A, B, C, D are given in Eq. (6.20), combining in a non-trivial way the imaginary parts of the quark self-energy induced from mesonic fluctuations derived in Section 5.2 and given in Eqs. (5.38) and (5.40).

---

[43] As it is seen empirically, the sub-dominant term $\sim D^{-1}$ contributes only $5 - 10\%$ but it stabilizes the rather involved numerics when evaluating the shear viscosity.

6.3. Results for the shear viscosity and the ratio η/s

We are now prepared to combine our findings and calculate the shear viscosity using the on-shell imaginary parts of Σ_j, $j = 0, 3, 4$, derived in Eqs. (5.38) and (5.40). They are used when evaluating the Kubo formula (6.26) numerically. The temperature dependence of the viscosity itself is shown in Fig. 6.2(a). Due to the on-shell conditions we have used when calculating Im Σ_j, only the temperature range above the Mott temperature, $T > T_M$, is accessible. One has $T_M(\mu = 0) = 212$ MeV and $T_M(\mu = 200 \text{ MeV}) = 171$ MeV. We observe an overall decreasing function $\eta(T)$ and also decreasing values for an increasing quark chemical potential, $\eta(\mu)$. As discussed in Section 4.2, a small shear viscosity displays a strongly correlated system: the larger the spectral width as a measure for the interaction strength, the lower the value of eta, cf. Fig. 4.2(a),(b). Although the viscosity given in Eq. (6.26) incorporates three independent imaginary parts, the effective spectral width $\tilde{\Gamma}$ shown in Fig. 5.4 already suggests this qualitative behavior. We conclude that the quark plasma described by the NJL model, where the shear viscosity is induced by mesonic fluctuations occurring at order $1/N_c$, becomes more strongly correlated for both increasing temperature and chemical potential. The overall scale of the ratio η/s is comparable to $1/4\pi$, but for large enough temperatures it undershoots the AdS/CFT benchmark. At vanishing chemical potential this happens at $T \approx 275$ MeV, at finite chemical potential, $\mu = 200$ MeV, even earlier at $T \approx 260$ MeV.

We compare our results with those from lattice QCD, [NS06, Mey07], which are shown as squares in Fig. 6.2(b). They have been derived within pure-gauge QCD and suggest a rising ratio $\eta/s(T)$ for $T > 200$ MeV. This behavior is not found in the NJL model. This qualitative difference can be explained by comparing to the results from hard thermal loop (HTL) calculation in QCD [AMY00, AMY03]. Both Abelian and non-Abelian gauge theories feature at leading-log

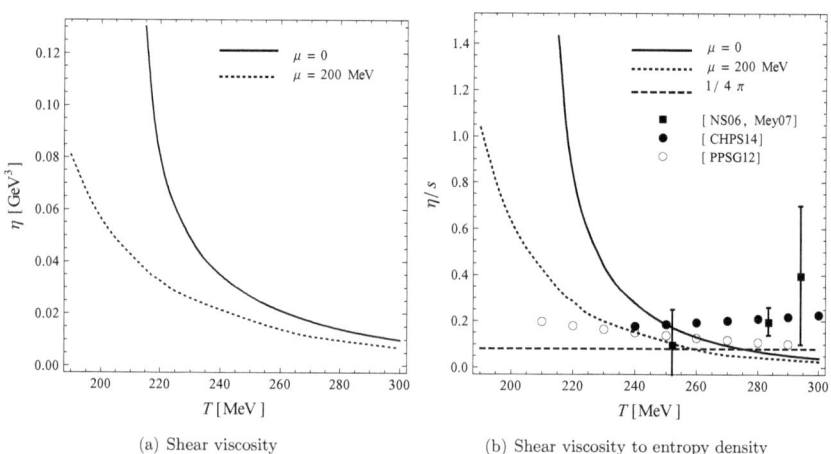

(a) Shear viscosity   (b) Shear viscosity to entropy density

Figure 6.2.: Temperature dependence of shear viscosity calculated from the NJL model in its large-N_c expansion for vanishing quark chemical potential and $\mu = 200$ MeV. See the discussion in the text.

order the behavior [KG06]:

$$\eta = \frac{\#_1 T^3}{\alpha_s^2 \ln(\#_2/\alpha_s)} \,, \tag{6.27}$$

hence $\eta/s \sim \alpha_s^{-2}$ at leading order. For increasing T, the strong coupling becomes weak, $\alpha_s \to 0$, therefore the ratio $\eta/s(T)$ rises when restricting to the HTL results. This trend is also seen (within uncertainties) in the lattice results already at rather small temperatures where HTL calculations are not applicable since they are based on perturbative-QCD and resummation techniques. However, the main reason for the rising behavior of the pure-gauge lattice results is asymptotic freedom and the relaxing correlation between the gauge bosons. In contrast, the NJL-model coupling G remains constant for increasing temperature. In addition, the magnitude of mesonic fluctuations are growing in the considered temperature range 180 MeV $\lesssim T \lesssim$ 300 MeV. In consequence, we find decreasing functions η and η/s. We also compare our results to those from [PPSG12], open circles in Fig. 6.2(b): the shear viscosity from the fundamental Kubo formula (4.1) has been evaluated numerically by calculating the cross section σ_{tot} from a parton cascade model with elastic two-body collisions using only gluonic degrees of freedom. Their results are described by

$$\frac{\eta}{s} = \frac{0.195}{\sigma_{\text{tot}} T^2} \,. \tag{6.28}$$

For the numerical evaluation in Fig. 6.2(b) we have used $\sigma_{\text{tot}} = 9\,\text{mb} = 0.9\,\text{fm}^2$. In comparison to our NJL results we observe a decreasing but flatter ratio η/s in this approach. The assumption of a temperature-independent total cross section does not describe the high-T behavior expected from HTL calculations and observed on the lattice. We also show this linear rise by the solid circles [CHPS14]. There, it is derived that this behavior can be described by

$$\left.\frac{\eta}{s}\right|_{\text{HTL}} = \frac{a}{\alpha_s^\gamma} \,, \tag{6.29}$$

with $a = 0.2$ and $\gamma = 1.6$ as their final fit for QCD. Note, that these fit parameters compare

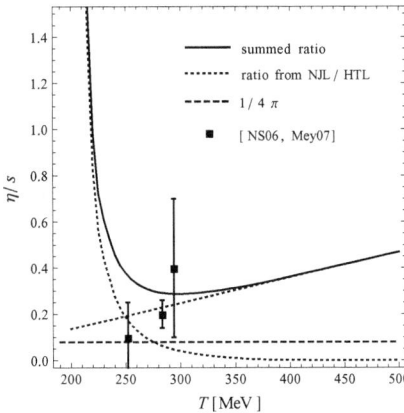

Figure 6.3.: Individual and summed ratios $\eta/s(T)$ from the NJL model (low-T region) and from HTL calculations (high-T region) at vanishing quark chemical potential. See the discussion in the text.

6. The ratio η/s in the NJL model

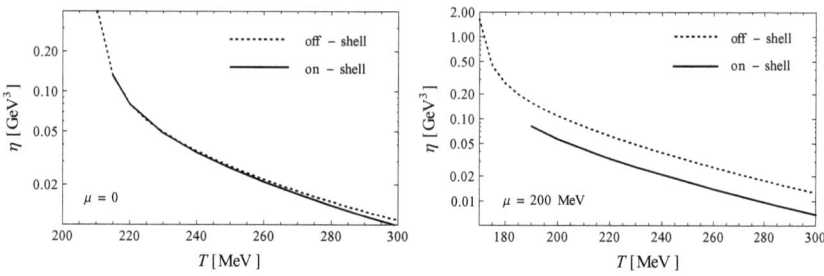

Figure 6.4.: Comparison between the on-shell and off-shell calculation of the shear viscosity at vanishing quark chemical potential (left panel) and $\mu = 200$ MeV (right panel)

well to the general gauge-theory results in Eq. (6.27). The analytical expression for the running QCD coupling uses a quasi-particle approach also in the vicinity of the (chiral) phase transition [Nes00]:

$$\alpha_{\rm s}(z(T)) = \frac{4\pi}{\beta_0} \frac{z^2 - 1}{z^2 \ln z^2}, \qquad (6.30)$$

with β_0 defined in Eq. (2.11), the reduced temperature $z = cT/T_{\rm c}$ with $T_{\rm c} \approx 155$ MeV and some fitted scale factor $c = 0.79$ in the QCD case [CHPS14]. The ratio η/s from HTL calculations is induced by dissipative processes in the gauge sector, where η/s calculated from the NJL model incorporates mesonic fluctuations in the quark sector. The sum of the two ratios[44] is shown in Fig. 6.3, where a minimum structure emerges due to the change between quarks and gluons as active degrees of freedom. It is located at $T_{\rm min} = 295$ MeV, where $\eta/s(T_{\rm min}) = 0.29$. In comparison to the results from [CHPS14], both the minimal value of η/s and its location are shifted to higher values: $T_{\rm min}^{\rm QCD} = 200$ MeV and $\eta/s(T_{\rm min}^{\rm QCD}) = 0.17$. The main reason for this shift is the rather large critical/crossover temperature, $T_{\rm c} = 190$ MeV, present in the two-flavor NJL model. However, the expectation for the temperature dependence of the ratio η/s as it is incorporated for hydrodynamic simulations, cf. the discussion related to Fig. 2.7, can be satisfied. Now, the ratio stays above the AdS/CFT benchmark: $\eta/s \gtrsim 3.5/4\pi$.

In Section 5.2.2 we have calculated additionally off-shell results for $\rho_j = \text{Im } \Sigma_j$, $j = 0, 3, 4$. They define the off-shell shear viscosity (6.26) in terms of Eqs. (6.20). The more involved numerical evaluation of the off-shell $\eta(T, \mu)$ using the results in Eqs. (5.67) and (5.68) are shown in Fig. 6.4. The main difference between the on-shell and off-shell is found on the qualitative level: there is no longer any restriction on the constituent-quark mass to provide a finite shear viscosity. It is interesting to observe that the small-T results smoothly extend the results fulfilling the Mott-condition $m_{\rm M} > 2m$ in the high-T region. At vanishing quark chemical potential the quantitative difference is almost negligible. At $\mu = 200$ MeV the difference is a simple factor shifting the viscosity to higher values, but the overall behavior of $\eta(T, \mu)$ is not changed.

We can explain this fact by going back to the parameter discussion in Section 4.2, where the peak-structure of the Kubo formula is examined in Fig. 4.3. We have seen that the main contribution to $\eta[\Gamma(p)]$ is collected around the integrand's pole position. Inspecting the Kubo formula for η incorporating the full Dirac structure of the quark propagator, Eq. (6.26), a similar structure is found: around the pole position where $D \to 0$, the main contributions to the integral

---
[44]Note that unlike the discussion at the very end of Section 2.3, in this case the contributions to the total shear viscosity come from different Kubo formulas for quarks and gluons. As we have discussed, different dissipative processes contributing to the same Kubo formula, however, do not sum up that simply.

6.3. Results for the shear viscosity and the ratio η/s

are collected. From Eqs. (6.19) and (6.20) one finds:

$$D \to 0 \quad \Leftrightarrow \quad N_1, N_2 \to 0 \quad \Rightarrow \quad p_0^2 - \boldsymbol{p}^2 - m^2 = 0 \ . \tag{6.31}$$

The last implication leads to the on-shell condition for the constituent quark. We conclude that the off-shell contributions are subleading effects due to the peak structure of the Kubo formula. It is remarkable that off-shell contributions do not strongly affect the shear viscosity despite the presence of additional imaginary parts. In the first instance one therefore expects the shear viscosity to become smaller. However, the distribution of the imaginary parts ρ_j, $j = 0, 3, 4$ in the integrand of η itself eventually leads to a rather small increase of the shear viscosity compared to the on-shell results.

7. Summary and Conclusion

"I am now convinced that theoretical physics is actually philosophy."[Moo92]

Max Born

In this thesis we have investigated the shear viscosity of a hot and dense quark plasma described by a large-N_c NJL model in the two-flavor case. In this non-perturbative model gluons are not treated as explicit, active degrees of freedom but they are hidden in effective NJL vertices that inherit all symmetries and the large-N_c scaling from QCD. We have used this scaling as bookkeeping method and have discussed the NJL model on this general footing. It has been reviewed how standard approaches from many-body physics can be derived: the gap equation defining thermal, dynamically generated constituent-quark masses, and the Bethe-Salpeter equation from which mesonic (soft) modes emerge. Our investigations have shown that the dominant attractive interaction channel for the evaluation of the shear viscosity is given by the quark-antiquark channel rather than the diquark channel, the latter being suppressed by its spectrum of large masses.

Transport coefficients such as the shear viscosity η are defined in systems out of thermodynamic equilibrium. In this thesis we have used the Kubo formalism which has been first reviewed and then investigated in a detailed parameter study. Assuming that the constituent-quark propagator can be described by a simple schematic parameterization of its spectral function, a suitable numerical approximation scheme was defined for the evaluation of the Kubo formula for shear viscosity. Its dependence on the shape of the spectral width and effects of thermal quark masses were investigated. Most importantly we have found a strong dependence of the shear viscosity on the NJL three-momentum cutoff which eventually ensures a meaningful scale of η. Whereas the three free NJL parameters (current-quark mass, NJL coupling strength and cutoff scale) are fixed by reproducing physical values for observables (meson masses, pion decay constant and chiral condensate), there is no model parameter left to adjust the overall scale of the shear viscosity. Therefore, our final results are parameter-free predictions within the large-N_c two-flavor NJL model. Going beyond the assumption of a simple one-width parameterization, a new Kubo formula for the shear viscosity has been derived, incorporating the full Dirac structure of the constituent-quark propagator.

In a large-N_c NJL model, the dominant dissipative process contributing to the shear viscosity is given by mesonic fluctuations. They are represented by virtual quark-antiquark loops resummed to all orders in the non-perturbative NJL coupling as it is described by the Bethe-Salpeter equation and treated as $1/N_c$ Fock contribution to the gap equation. We have calculated the three Dirac self-energy contributions given by this Fock term using both on-shell and off-shell conditions. Evaluating the new Kubo formula using these results we have found a decreasing shear viscosity as function of both temperature and quark chemical potential. At vanishing chemical potential, off-shell effects extend the on-shell results also to the low-temperature region where the on-shell phase space is collapsing. Apart from this, off-shell effects have no further quantitative or qualitative influence. However, at finite quark chemical potentials, off-shell effects shift the shear viscosity to higher values but its overall qualitative behavior is not affected.

We have observed that the dimensionless ratio η/s undershoots the AdS/CFT benchmark $1/4\pi$ at large enough temperatures. This statement is, however, true only as long as we stick to

7. Summary and Conclusion

quark degrees of freedom. Combining our result for the shear viscosity with perturbative results from hard-thermal-loop calculations in the high-T region, we find the ratio η/s to feature a minimum structure well above the AdS/CFT benchmark. The combined results compare well with those from lattice QCD setting the overall behavior and scale of the ratio η/s. However, the crossover temperature in the two-flavor NJL model, $T_c \approx 190$ MeV, is larger than the lattice value, $T_c \approx 155$ MeV. Therefore, the onset of the dominant on-shell contributions to the shear viscosity is shifted to larger temperatures.

Since the mid 1970s mesonic fluctuations (the meson cloud) have been known as significant contributions to the nucleon mass. In chiral (cloudy) bag models they induce a correction term lowering the leading-order nucleon mass. In contrast, we have found a positive correction from the pionic fluctuation to the self-energy of the constituent quark. This qualitatively different behavior is understood by expanding our field-theoretical calculation in the non-relativistic limit appropriate for comparison with the bag-model results. The sign flip is explained by an over-compensation of the non-relativistic contributions by purely relativistic terms. The comparison of our results with those from the well-known chiral bag models has served as an instructive and important cross check of our calculations.

It has been known that using the Kubo formalism for calculating transport coefficients is a rather inefficient approach since already at leading order in a weakly coupled theory resummation techniques have to be applied. The NJL model is non-perturbative by construction, hence our evaluation of the shear viscosity is such as well. We have concluded that ladder-diagram resummation in the Kubo sector is only sub-dominant and therefore not necessary.

One of the main results of this thesis is in fact the derivation of a new Kubo formula that takes the (non-perturbative) Dirac structure of the relativistic quark propagator fully into account. It has been evaluated for both on-shell and off-shell quark spectral functions in a consistent way.

In summary, we have found within the two-flavor large-N_c NJL model an overall decreasing shear viscosity $\eta(T,\mu)$ which undershoots the AdS/CFT benchmark at sufficiently large temperatures. The underlying dissipative process is given by mesonic fluctuations in the quark sector arising as Fock term in the gap equation. Comparing on-shell and off-shell contributions to the shear viscosity shows no significant difference on the final results which have been derived without resumming ladder diagrams in the Kubo sector. Combining the NJL results for the shear viscosity with results from hard-thermal-loop calculations leads to a minimum structure of η/s located above the AdS/CFT benchmark. The correlated quark plasma described in this thesis features a small shear viscosity characteristic of a perfect fluid.

A. Appendix

> "This book – I mean the universe – is written in
> the mathematical language."[Bur03]
>
> Galileo Galilei

A.1. Group-theoretic details of $\mathrm{SU}(N)$

Let $G = \mathrm{SU}(N)$ denote the Lie group and $\mathcal{G} = \mathrm{su}(N)$ its Lie algebra with generators T_a, $a \in \{1, \ldots, N^2 - 1\}$. They are traceless and Hermitian. We also define $\lambda_a = 2T_a$. The Lie bracket, $[\cdot, \cdot] : \mathcal{G} \times \mathcal{G} \to \mathcal{G}$, provides the multiplication on the algebra in terms of the *structure constants*, f_{abc}:

$$[T_a, T_b] = \mathrm{i} f_{abc} T_c , \qquad [\lambda_a, \lambda_b] = 2\mathrm{i} f_{abc} \lambda_c . \tag{A.1}$$

Note that the commutator of two generators is skew Hermitian (and traceless), hence one can write it as a linear combination of $\mathrm{i}\lambda_c$. In addition, the structure constants can be real numbers only[45]: $f_{abc} \in \mathbb{R}$.

Since G is a matrix Lie group, there is also the usual matrix multiplication available. One can decompose the product $\lambda_a \lambda_b$ into a Hermitian and an skew-Hermitian part,

$$\lambda_a \lambda_b = \frac{1}{2}[\lambda_a, \lambda_b] + \frac{1}{2}\{\lambda_a, \lambda_b\} = i f_{abc} \lambda_c + \xi(r) \delta_{ab} \mathbb{1}_{n \times n} + d_{abc} \lambda_c , \tag{A.2}$$

parameterized by the structure constants f_{abc} (totally antisymmetric) and d_{abc} (totally symmetric), and the representation dependent $\xi(r)$. n denotes the dimension of the representation. The anti-commutator of two generators is Hermitian, but in general not traceless. Note that any Hermitian $n \times n$ matrix, A, can be decomposed into a Hermitian, traceless matrix and a diagonal matrix:

$$A = \left(A - \frac{\mathrm{tr}\, A}{n} \mathbb{1}_{n \times n} \right) + \frac{\mathrm{tr}\, A}{n} \mathbb{1}_{n \times n} . \tag{A.3}$$

We also introduce the two representation-dependent quantities, $C(\mathrm{r})$ and $C_2(\mathrm{r})$, called the *Dynkin index* and *quadratic Casimir operator*, respectively.

$$\mathrm{tr}\, T_a^{\mathrm{r}} T_b^{\mathrm{r}} =: C(\mathrm{r}) \delta_{ab} , \qquad T_a^{\mathrm{r}} T_a^{\mathrm{r}} =: C_2(\mathrm{r}) \mathbb{1}_{n \times n} . \tag{A.4}$$

For SU(2) we have in fundamental representation the usual Pauli matrices[46] $\lambda_a = \sigma_a$ with $[\sigma_a, \sigma_b] = 2\mathrm{i}\epsilon_{abc}\sigma_c$ ($a, b, c = 1, 2, 3$). They are given by

$$\sigma_1 := \begin{pmatrix} 0 & 1 \\ 1 & 0 \end{pmatrix}, \quad \sigma_2 := \begin{pmatrix} 0 & -\mathrm{i} \\ \mathrm{i} & 0 \end{pmatrix}, \quad \sigma_3 := \begin{pmatrix} 1 & 0 \\ 0 & -1 \end{pmatrix}, \quad \boldsymbol{\sigma} := (\sigma_1, \sigma_2, \sigma_3) , \tag{A.5}$$

and $\sigma_0 := \mathbb{1}$, with $\mathrm{tr}\, (\sigma_a \sigma_b) = 2\delta_{ab}$ ($a, b = 0, 1, 2, 3$). One has $f_{abc} = \epsilon_{abc}$ and $d_{abc} = 0$.

---
[45] From this it follows immediately that the adjoint representation, A, is a real representation for all SU(N).
[46] The Pauli matrices describe both spin and isospin and are denoted by σ_a and τ_a, respectively.

A. Appendix

For SU(3) we have in fundamental representation the usual Gell-Mann matrices with $[\lambda_a, \lambda_b] = 2\mathrm{i} f_{abc}\lambda_c$ ($a, b, c = 1, \ldots, 8$). They are given by

$$\lambda_1 = \begin{pmatrix} 0 & 1 & 0 \\ 1 & 0 & 0 \\ 0 & 0 & 0 \end{pmatrix}, \quad \lambda_2 = \begin{pmatrix} 0 & -\mathrm{i} & 0 \\ \mathrm{i} & 0 & 0 \\ 0 & 0 & 0 \end{pmatrix}, \quad \lambda_3 = \begin{pmatrix} 1 & 0 & 0 \\ 0 & -1 & 0 \\ 0 & 0 & 0 \end{pmatrix},$$

$$\lambda_4 = \begin{pmatrix} 0 & 0 & 1 \\ 0 & 0 & 0 \\ 1 & 0 & 0 \end{pmatrix}, \quad \lambda_5 = \begin{pmatrix} 0 & 0 & -\mathrm{i} \\ 0 & 0 & 0 \\ \mathrm{i} & 0 & 0 \end{pmatrix}, \quad \lambda_6 = \begin{pmatrix} 0 & 0 & 0 \\ 0 & 0 & 1 \\ 0 & 1 & 0 \end{pmatrix}, \quad (\mathrm{A}.6)$$

$$\lambda_7 = \begin{pmatrix} 0 & 0 & 0 \\ 0 & 0 & -\mathrm{i} \\ 0 & \mathrm{i} & 0 \end{pmatrix}, \quad \lambda_8 = \frac{1}{\sqrt{3}}\begin{pmatrix} 1 & 0 & 0 \\ 0 & 1 & 0 \\ 0 & 0 & -2 \end{pmatrix}, \quad \lambda_0 = \sqrt{\frac{2}{3}}\begin{pmatrix} 1 & 0 & 0 \\ 0 & 1 & 0 \\ 0 & 0 & 1 \end{pmatrix}.$$

One has

$$f_{123} = 1, \quad f_{147} = f_{165} = f_{246} = f_{257} = f_{345} = f_{376} = \frac{1}{2}, \quad f_{458} = f_{678} = \frac{\sqrt{3}}{2}, \quad (\mathrm{A}.7)$$

$$d_{118} = d_{228} = d_{338} = -2d_{448} = -2d_{558} = -2d_{668} = -2d_{778} = -d_{888} = \frac{1}{\sqrt{3}},$$

$$d_{146} = d_{157} = -d_{247} = d_{256} = d_{344} = d_{355} = -d_{366} = -d_{377} = \frac{1}{2}. \quad (\mathrm{A}.8)$$

We summarize for all $N \in \mathbb{N}$ the most important properties of SU(N) representations in the following Table A.1, where we denote by $n = \dim(\mathrm{r})$ the dimension of the representation:

r	n	C	C_2	ξ
F	N	$\frac{1}{2}$	$\frac{N^2-1}{2N}$	$\frac{2}{N}$
A	$N^2 - 1$	N	N	$\frac{4N}{N^2-1}$
$\bar{\mathrm{r}}_s$	$\frac{N(N+1)}{2}$	$\frac{N+2}{2}$	$\frac{(N+2)(N-1)}{N}$	$\frac{2N}{N-1}$
$\bar{\mathrm{r}}_a$	$\frac{N(N-1)}{2}$	$\frac{N-2}{2}$	$\frac{(N+1)(N-2)}{2}$	$\frac{2N}{N+1}$

Table A.1.: Properties of the most important SU(N) representations

The Dynkin index and the quadratic Casimir operator are not independent quantities, but they are constrained by

$$(N^2 - 1)C(\mathrm{r}) = n\, C_2(\mathrm{r}), \quad (\mathrm{A}.9)$$

which follows immediately from their definitions (A.4). One also calculates

$$\xi(\mathrm{r}) = \frac{4\, C_2(\mathrm{r})}{N^2 - 1}. \quad (\mathrm{A}.10)$$

We remark the following: f_{abc} is totally antisymmetric, i.e. $f_{abc} = f_{bca} = f_{cab} = -f_{bac}$, whereas d_{abc} is totally symmetric, i.e. $d_{abc} = d_{bca} = d_{cab} = +d_{bac}$. In total, there are 54 and 58 non-vanishing entries in f_{abc} and d_{abc}, respectively.

A.1. Group-theoretic details of SU(N)

From Eq. (A.2) it follows immediately that

$$\{\lambda_a, \lambda_b\} = 2\xi(r)\delta_{ab}\mathbb{1}_{n\times n} + 2d_{abc}\lambda_c \stackrel{\text{F}}{=} \frac{4}{N}\delta_{ab}\mathbb{1}_{n\times n} + 2d_{abc}\lambda_c \;, \tag{A.11}$$

where the last identity is valid for the fundamental representation only. In this representation one defines also $\lambda_0 := \sqrt{\xi}\,\mathbb{1}_{N\times N}$, such that the following holds for all $a, b \in \{0, 1, \ldots, N\}$:

$$\operatorname{tr}\lambda_a\lambda_b = 2\delta_{ab} \;. \tag{A.12}$$

A. Appendix

A.2. Matsubara formalism

We summarize briefly the most important facts about the Matsubara formalism we have used in this work. An introduction to this imaginary time formalism can be found in standard textbooks, e.g. [LB00, KG06]. In Matsubara space the (imaginary) time coordinate is Wick rotated and allows only for discrete frequencies: $\tau = -it \in [0,\beta] \subset \mathbb{R}$, where we denote the inverse temperature by $\beta = 1/T$. Periodic (anti-periodic) boundary conditions for bosons (fermions) at $\tau = \beta$ lead to a discrete but infinite set of Matsubara frequencies: with $n \in \mathbb{Z}$ one finds for bosons $\omega_n = 2n\pi T - i\mu$, whereas they read for fermions $\nu_n = (2n+1)\pi T - i\mu$. Carrying the discrete sums over Matsubara frequencies leads to Bose and Fermi distributions which are defined as

$$n_B(E \mp \mu) = n_B^{\pm}(E) = \frac{1}{e^{\beta(E \mp \mu)} - 1}, \quad n_F(E \mp \mu) = n_F^{\pm}(E) = \frac{1}{e^{\beta(E \mp \mu)} + 1}. \quad (A.13)$$

A.2.1. Review of propagators

The bosonic and fermionic propagators in Matsubara space are summarized and compared to those in Euclidean and Minkowskian spacetime.

Propagators for bosons

Thermal (Matsubara) propagator

$$G_\beta^B(\boldsymbol{p}, \omega_n) = \frac{1}{\omega_n^2 + \boldsymbol{p}^2 + m^2}. \quad (A.14)$$

Propagator in Euclidean spacetime

$$G_E^B(\boldsymbol{p}, p_4) = \frac{1}{p_4^2 + \boldsymbol{p}^2 + m^2} = \frac{1}{p_E^2 + m^2}. \quad (A.15)$$

Propagator in Minkowski spacetime

$$G_M^B(p_0, \boldsymbol{p}) = \frac{1}{p_0^2 - \boldsymbol{p}^2 - m^2} = \frac{1}{p^2 - m^2}. \quad (A.16)$$

The following is true: $G_\beta^B(\boldsymbol{p}, -p_4) = G_E^B(\boldsymbol{p}, p_4)$ and $G_E^B(\boldsymbol{p}, ip_0) = -G_M^B(p_0, \boldsymbol{p})$. We have used

$$p_4 = ip_0, \ p_0 = i\omega_n \ \Leftrightarrow \ p_4 = -\omega_n.$$

Propagators for fermions

Thermal (Matsubara) propagator:

$$G_\beta^F(\boldsymbol{p}, \nu_n) = \frac{1}{-\nu_n \gamma_4 + \boldsymbol{p} \cdot \boldsymbol{\gamma} + m} = \frac{\nu_n \gamma_4 - \boldsymbol{p} \cdot \boldsymbol{\gamma} + m}{\nu_n^2 + \boldsymbol{p}^2 + m^2}. \quad (A.17)$$

Propagator in Euclidean spacetime

$$G_E^F(\boldsymbol{p}, p_4) = -\frac{\slashed{p}_E - m}{p_E^2 + m^2} = -\frac{p_4 \gamma_4 + \boldsymbol{p} \cdot \boldsymbol{\gamma} - m}{p_4^2 + \boldsymbol{p}^2 + m^2}. \quad (A.18)$$

Propagator in Minkowskian spacetime

$$G_\mathrm{M}^\mathrm{F}(p_0, \boldsymbol{p}) = \frac{1}{\not{p} - m} = \frac{\not{p} + m}{p^2 - m^2} = \frac{p_0\gamma_0 - \boldsymbol{p}\cdot\boldsymbol{\gamma} + m}{p_0^2 - \boldsymbol{p}^2 - m^2} \,. \tag{A.19}$$

The following is true: $G_\beta^\mathrm{F}(\boldsymbol{p}, -p_4) = G_\mathrm{E}^\mathrm{F}(\boldsymbol{p}, p_4)$ and $G_\mathrm{E}^\mathrm{F}(\boldsymbol{p}, \mathrm{i}p_0) = -G_\mathrm{M}^\mathrm{F}(p_0, \boldsymbol{p})$. We have used for $\mu, \nu \in \{0,1,2,3\}$ and $i,j \in \{1,2,3,4\}$:

$$p_4 = \mathrm{i}p_0\,,\ p_0 = \mathrm{i}\nu_n\,,\ \gamma_4 = \mathrm{i}\gamma_0\,,\ \{\gamma_\mu, \gamma_\nu\} = 2g_{\mu\nu}\mathbf{1}\,,\ \{\gamma_i, \gamma_j\} = -2\delta_{ij}\mathbf{1}\,.$$

A.2.2. Master formulas

For the bosonic and fermionic Matsubara frequencies, $\omega_n = 2n\pi T - \mathrm{i}\mu$ and $\nu_n = (2n+1)\pi T - \mathrm{i}\mu$, respectively. The frequencies for the antiparticles are denoted by $\bar\omega_n = 2n\pi T + \mathrm{i}\mu = \omega_n^*$, and $\bar\nu_n = (2n+1)\pi T + \mathrm{i}\mu = \nu_n^*$, respectively. One has:

$$T\sum_{n\in\mathbb{Z}} \frac{1}{\omega_n^2 + \omega^2} = \frac{1}{\omega}\left(\frac{1}{2} + n_\mathrm{B}(\omega)\right) = \frac{1}{4\omega}\left(\coth\left(\frac{\omega - \mu}{2T}\right) + \coth\left(\frac{\omega + \mu}{2T}\right)\right) \tag{A.20}$$

$$T\sum_{n\in\mathbb{Z}} \frac{1}{\nu_n^2 + \omega^2} = \frac{1}{\omega}\left(\frac{1}{2} - n_\mathrm{F}(\omega)\right) = \frac{1}{4\omega}\left(\tanh\left(\frac{\omega - \mu}{2T}\right) + \tanh\left(\frac{\omega + \mu}{2T}\right)\right) \tag{A.21}$$

The $\coth(\cdot)$ and $\tanh(\cdot)$ describe Bose and Fermi distributions:

$$\coth\left(\frac{E \mp \mu}{2T}\right) = 1 + 2n_\mathrm{B}(E \mp \mu) = 1 + 2n_\mathrm{B}^\pm(E)\,, \tag{A.22}$$

$$\tanh\left(\frac{E \mp \mu}{2T}\right) = 1 - 2n_\mathrm{F}(E \mp \mu) = 1 - 2n_\mathrm{F}^\pm(E)\,. \tag{A.23}$$

These general building blocks are useful to carry out the Matsubara frequencies when calculating the quark self-energy from mesonic fluctuations in Section 5.2. For the evaluation of Σ_0 and Σ_3 in Eq. (5.8) we have used:

$$\begin{aligned}
T\sum_{n\in\mathbb{Z}} & \frac{1}{[(\nu - \nu_n)^2 + E_b^2][\nu_n^2 + E_f^2]} = \\
&= \frac{1}{2E_bE_f[(E_b + E_f)^2 + \nu^2][(E_b - E_f)^2 + \nu^2]}\left[E_f(\nu^2 + E_f^2 - E_b^2)\coth\left(\frac{E_b}{2T}\right) + \right.\\
&\quad \left. + E_b(\nu^2 + E_b^2 - E_f^2)\cdot\frac{1}{2}\left(\tanh\left(\frac{E_f - \mu}{2T}\right) + \tanh\left(\frac{E_f + \mu}{2T}\right)\right)\right] + \\
&\quad - \frac{\mathrm{i}\nu}{[(E_b + E_f)^2 + \nu^2][(E_b - E_f)^2 + \nu^2]}\cdot\frac{1}{2}\left(\tanh\left(\frac{E_f - \mu}{2T}\right) - \tanh\left(\frac{E_f + \mu}{2T}\right)\right) = \\
&= \frac{1}{2E_bE_f}\left[\frac{(E_f + E_b)\left[1 + n_\mathrm{B}(E_b) - \frac{1}{2}\left(n_\mathrm{F}^+(E_f) + n_\mathrm{F}^-(E_f)\right)\right]}{(E_f + E_b)^2 + \nu^2} + \right.\\
&\quad \left. + \frac{(E_f - E_b)\left[n_\mathrm{B}(E_b) + \frac{1}{2}\left(n_\mathrm{F}^+(E_f) + n_\mathrm{F}^-(E_f)\right)\right]}{(E_f - E_b)^2 + \nu^2}\right] + \\
&\quad + \frac{\mathrm{i}\nu\left[n_\mathrm{F}^+(E_f) - n_\mathrm{F}^-(E_f)\right]}{[(E_f + E_b)^2 + \nu^2][(E_f - E_b)^2 + \nu^2]}\,.
\end{aligned} \tag{A.24}$$

A. Appendix

For the evaluation of Σ_4 in Eq. (5.8) we have used:

$$T\sum_{n\in\mathbb{Z}} \frac{\nu_n}{[(\nu-\nu_n)^2 + E_b^2][\nu_n^2 + E_f^2]} =$$

$$= \frac{\nu}{2E_b[(E_b+E_f)^2 + \nu^2][(E_b-E_f)^2 + \nu^2]} \left[(E_b^2 + E_f^2 + \nu^2) \coth\left(\frac{E_b}{2T}\right) + \right.$$

$$\left. -2E_b E_f \cdot \frac{1}{2}\left(\tanh\left(\frac{E_f - \mu}{2T}\right) + \tanh\left(\frac{E_f + \mu}{2T}\right)\right)\right] +$$

$$- \frac{E_b^2 - E_f^2 + \nu^2}{2\mathrm{i}\,[(E_b+E_f)^2 + \nu^2][(E_b-E_f)^2 + \nu^2]} \cdot \frac{1}{2}\left(\tanh\left(\frac{E_f - \mu}{2T}\right) - \tanh\left(\frac{E_f + \mu}{2T}\right)\right) =$$

$$= \frac{\nu}{2E_b}\left[\frac{1 + n_\mathrm{B}(E_b) - \frac{1}{2}\left(n_\mathrm{F}^+(E_f) + n_\mathrm{F}^-(E_f)\right)}{(E_f + E_b)^2 + \nu^2} + \frac{n_\mathrm{B}(E_b) + \frac{1}{2}\left(n_\mathrm{F}^+(E_f) + n_\mathrm{F}^-(E_f)\right)}{(E_f - E_b)^2 + \nu^2}\right] +$$

$$- \frac{(E_b^2 - E_f^2 + \nu^2)\left[n_\mathrm{F}^+(E_f) - n_\mathrm{F}^-(E_f)\right]}{2\mathrm{i}\,[(E_f + E_b)^2 + \nu^2][(E_f - E_b)^2 + \nu^2]}.$$

(A.25)

For evaluating the Matsubara sums the following identity has been useful:

$$\frac{\sinh(x)}{\cosh(x) + \cosh(y)} = \frac{1}{2}\left[\tanh\left(\frac{x-y}{2}\right) + \tanh\left(\frac{x+y}{2}\right)\right]. \tag{A.26}$$

A.3. Spectral representation of propagators

In this brief appendix we review the Källén-Lehmann spectral representation for retarded and advanced propagators in Minkowski and also Matsubara space. Thereby we fix first our sign convention for the analytical continuation between these two spaces:

$$[\cdot]_\beta \leftrightarrow -[\cdot]_{R/A} \quad \text{for} \quad i[\cdot]_n \leftrightarrow p_0 \pm i\epsilon . \tag{A.27}$$

For evaluating integrals containing a first-order pole on the real axis, we interpret it as principal value integral following the Sokhotski-Plemelj Theorem:

$$\int_\mathbb{R} dx \, \frac{f(x)}{(x-a) \pm i\epsilon} = \mp i\pi f(a) + \fint_\mathbb{R} dx \, \frac{f(x)}{x-a} , \tag{A.28}$$

where the principal value integral is defined by cutting out a symmetric interval around the pole position $B_\epsilon(a) = \{x \in \mathbb{R} : |x-a| < \epsilon\}$:

$$\fint_\mathbb{R} dx \, \frac{f(x)}{x-a} = \lim_{\epsilon \to 0} \int_{\mathbb{R} \setminus B_\epsilon(a)} dx \, \frac{f(x)}{x-a} . \tag{A.29}$$

If the integrand is regular on the real axes, $f(a) = 0$, then the principal value integral reduces to the common integral. The validity of the relation (A.28) is crucially related to the integration range $X = \mathbb{R} \ni a$, since one uses

$$\lim_{\epsilon \to 0} \frac{\epsilon}{(x-a)^2 + \epsilon^2} = -\lim_{\epsilon \to 0} \text{Im} \, \frac{1}{(x-a) + i\epsilon} = \pi \delta(x-a) . \tag{A.30}$$

If the pole lies outside the integration range, $a \notin X$, then the delta function does not contribute and again the principal value integral reduces to the common integral.

The spectral function $\rho(\omega, \mathbf{p})$ which relates to the Minkowskian propagators is defined by the integral identity

$$G_{R/A}(p_0, \mathbf{p}) = \int_{-\infty}^{\infty} d\omega \, \frac{\rho(\omega, \mathbf{p})}{p_0 - \omega \pm i\epsilon} . \tag{A.31}$$

Note that $G_R(p)$ denotes just the Feynman propagator describing a retarded and advanced propagation of particles and antiparticles, respectively. We determine the spectral function in such a way that the resulting propagators for retarded and advanced fermions and bosons have the expected form:

$$\begin{aligned}
G_{R/A}^B &= \frac{1}{p_0^2 - \mathbf{p}^2 - m^2 \pm i\,\text{sgn}(p_0)\epsilon} = \frac{1}{p^2 - m^2 \pm i\,\text{sgn}(p_0)\epsilon} , \\
G_{R/A}^F &= \frac{p_0 \gamma_0 - \mathbf{p} \cdot \boldsymbol{\gamma} + m}{p_0^2 - \mathbf{p}^2 - m^2 \pm i\,\text{sgn}(p_0)\epsilon} = \frac{1}{\slashed{p} - m \pm i\,\text{sgn}(p_0)\epsilon} .
\end{aligned} \tag{A.32}$$

The free spectral functions $\rho_0(p_0, \mathbf{p})$ for bosons and fermions read with $E = \sqrt{\mathbf{p}^2 + m^2}$:

$$\begin{aligned}
\rho_0^B &= \text{sgn}(p_0) \, \delta(p_0^2 - E^2) , \\
\rho_0^F &= \frac{p_0 \gamma_0 - \mathbf{p} \cdot \boldsymbol{\gamma} + m}{2p_0} \delta(|p_0| - E) .
\end{aligned} \tag{A.33}$$

Using the principal-value integral one derives

$$\lim_{\epsilon \to 0^+} \text{Im} \, G_{R/A}(p) = \mp \pi \rho(p) . \tag{A.34}$$

A. Appendix

The spectral function is related to the highly singular behavior of the imaginary part of Minkowski propagators:

$$\lim_{\epsilon \to 0} \operatorname{Im} \frac{a + ib\epsilon}{x + i\epsilon} = -a\pi\delta(x) , \qquad (A.35)$$

where $a, b, x \in \mathbb{R}$ has been assumed. Using this identity, our claim for ρ_0^B and ρ_0^F follows.
In general, the following symmetry relations hold:

$$\begin{aligned}
\rho^B(-\omega) &= -\rho^B(\omega) , \\
\operatorname{tr}(\gamma_0 \rho^F(-\omega)) &= +\operatorname{tr}(\gamma_0 \rho^F(\omega)) , \\
\operatorname{tr}(\boldsymbol{\gamma} \rho^F(-\omega)) &= -\operatorname{tr}(\boldsymbol{\gamma} \rho^F(\omega)) , \\
\operatorname{tr}(\rho^F(-\omega)) &= -\operatorname{tr}(\rho^F(\omega)) .
\end{aligned} \qquad (A.36)$$

For the propagators in Matsubara space we define

$$G_\beta(\boldsymbol{p}, \omega_n) = \int_{-\infty}^{\infty} d\omega \frac{\rho(\omega, \boldsymbol{p})}{\omega - i\omega_n} . \qquad (A.37)$$

With this definition, the free spectral functions lead to the Matsubara propagators as we expect:

$$G_\beta(\boldsymbol{p}, \omega_n)\big|_{\rho=\rho_0^B} = \frac{1}{\omega_n^2 + \boldsymbol{p}^2 + m^2} , \quad \text{and} \quad G_\beta(\boldsymbol{p}, \nu_n)\big|_{\rho=\rho_0^F} = \frac{\nu_n \gamma_4 - \boldsymbol{p}\boldsymbol{\gamma} + m}{\nu_n^2 + \boldsymbol{p}^2 + m^2} . \qquad (A.38)$$

For the analytical continuation to Minkowski space we find indeed the (negative) propagator, including naturally the $\operatorname{sgn}(p_0)$ term in the denominator:

$$G_\beta^{B/F}\bigg|_{i[\cdot]_n = p_0 \pm i\epsilon} \mapsto -G_{R/A}^{B/F} . \qquad (A.39)$$

A.4. Analytical properties of the quark-meson coupling

We study the analytical structure of principal-value integrals where external conditions (such as temperature, quark chemical potential or meson energy) determine the position of the pole in the complex integration plane. In particular we study the questions what happens if the pole enters the interval of integration and the integration range starts or ends exactly at the pole position. This rather technical question is related to the temperature dependence of the quark-meson coupling having not fixed the meson mass in the first place (cf. the discussion in Section 3.5, particularly the comparison between Eq. (3.104) and Eq. (3.117)). We have wondered about the divergent derivative of I_2 compared to the convergent (but kinky) structure of I_2; this appears to be notable, since both integrals share the same pole structure and they are both interpreted as principal-value integrals.

As prototype of principal-values integral we consider for some $n \in \mathbb{N}$ and $a > 0$ the following set of functions:
$$f_a^{(n)}(q) = \frac{q^n}{q^2 - (a-1)}, \qquad (A.40)$$

and its integral
$$F_a^{(n)} = \int_0^1 dq\, f_a^{(n)}(q)\,. \qquad (A.41)$$

For $a < 1$ or $a > 2$ the pole $q_0 = \sqrt{a-1} \in \mathbb{C}$ lies offside the integration interval $q \in [0,1]$, compare Fig. A.1. For $a = 0$ the pole is $q = \pm i$; for increasing a it approaches the origin along the imaginary axis. For $1 \leq a \leq 2$ the pole is located in the integration interval and $F_a^{(n)}$ is interpreted as principal-value integral. We find for $n = 0$ in the different regions:

$$F_a^{(0)} = \begin{cases} \frac{1}{\sqrt{1-a}} \mathrm{ArcCot}(\sqrt{1-a}) & \text{for } a < 1\,, \\[4pt] \frac{1}{2\sqrt{a-1}} \log\left(\frac{2\sqrt{a-1}-a}{a-2}\right) & \text{for } 1 < a < 2\,, \\[4pt] \frac{1}{2\sqrt{a-1}} \log\left(\frac{2\sqrt{a-1}-a}{2-a}\right) & \text{for } a > 2\,. \end{cases} \qquad (A.42)$$

Having a closer look at the boundaries of the integration range, $a = 1$ and $a = 2$, one finds the

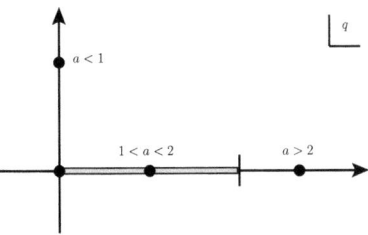

Figure A.1.: Influence of the external parameter a on the pole position and the range of integration (grey area) in $F_a^{(n)}$

$F_a^{(n)}$	$n=0$	$n=1$	$n=2$
$\lim_{a \to 1-}$	∞	∞	1
$\lim_{a \to 1+}$	-1	∞	1
$\lim_{a \to 2-}$	$-\infty$	$-\infty$	$-\infty$
$\lim_{a \to 2+}$	$-\infty$	$-\infty$	$-\infty$

Table A.2.: Behavior of principal-value integrals when the integration starts in the pole position

A. Appendix

following limits:
$$\lim_{a \to 1^-} F_a^{(0)} = \infty, \quad \lim_{a \to 1^+} F_a^{(0)} = -1 \tag{A.43}$$

$$\lim_{a \to 2^-} F_a^{(0)} = -\infty, \quad \lim_{a \to 1^+} F_a^{(0)} = -\infty \tag{A.44}$$

We now vary additionally $n \in \{0, 1, 2, \ldots\}$ in order to smooth out the pole structure for $q \to 0$. Our findings are summarized in Table A.2. The implication of this study to the physical behavior of the meson masses and the quark-meson coupling is far reaching: the meson masses are determined by Eq. (3.97) that has a factor q^2 from the Jacobian. This refers to the case $n = 2$ and therefore their thermal dependencies feature just a kink. In contrast, the quark-meson coupling in Eq. (3.117) has been derived after once integrating by parts with respect to q, leading to an expression referring to the case $n = 0$. Now, the integral becomes divergent when the pole approaches the origin.

A.5. List of symbols

a_s	reduced strong fine-structure constant		
α_s	strong fine-structure constant	N_c	number of colors
A_a^μ	axialvector current of SU(N)	N_f	number of flavor
β	inverse temperature	ν_n	fermionic Matsubara frequencies
	or beta function of QCD	$|0\rangle$	perturbative vacuum
β_s	inverse proper temperature	$|\Omega\rangle$	non-perturbative vacuum
$c_n\{2k\}$	$2k$-particle cumulant	$\Omega^\text{S/P}$	fermionic-pole correction
c_m^k, c_{mn}^{kl}	Fierz coefficients	P	pressure
δ_n	non-flow contributions to c_n	$\mathcal{P}$	path-ordering symbol
D_μ	covariant derivative	$\pi_{\mu\nu}$	viscous-stress (shear) tensor
$\Delta_{\mu\nu}$	projector	$\Pi^\text{S/P}$	polarization tensor
ϵ	energy density	ϕ	Boson field
ϵ_n	spatial anisotropies	Φ	2PI generating functional
η	shear viscosity		*or* renormalized Polyakov loop
f_π	pion decay constant	Ψ_RP	reaction-plane angle
$\mathcal{F}$	Fierz-transformation operator	$\langle \bar{\psi}\psi \rangle$	chiral condensate
	or free energy	$\langle :\bar{\psi}\psi: \rangle$	subtracted chiral condensate
g	closed string coupling	$\langle \bar{q}q \rangle$	quark-antiquark condensate
$g_{\pi qq}$	quark-pion coupling	Q_5	axialvector charge of U(1)
g_Mqq	quark-meson coupling	Q_a^A	axialvector charge of SU(N)
g_QCD	fundamental QCD coupling	Q_a^V	vector charge of SU(N)
G	NJL four-fermion vertex	R	curvature radius in AdS$_5$ space
$\mathcal{G}^\pm$	off-shell auxiliary functions	s	entropy density
$G^{\mu\nu}$	gluonic field-strength tensor	ρ	non-equilibrium statistical operator
$G_\beta^\text{B/F}$	propagators in Matsubara space		*or* spectral function
$G_\text{M,R/A}^\text{B/F}$	propagators in Minkowski space	ρ_0	equilibrium statistical operator
H	NJL $2N_\text{f}$ vertex	ρ_0, ρ_3, ρ_4	imaginary parts of quark self-energies
$\mathcal{H}$	on-shell auxiliary function	s	entropy density
Γ	spectral width	$\Sigma^\text{S/P}(p^2)$	vacuum quark self-energy
	or Lorentz structure	$\Sigma_\beta^\text{S/P}(\boldsymbol{p},\nu_n)$	thermal quark self-energy
$\tilde{\Gamma}$	effective spectral width	$\sqrt{\sigma}$	string tension
j_5^μ	axialvector current of U(1)	T^a	infinitesimal generators
$J^\text{I}, J^\text{II}, J^\text{III}$	off-shell contributions to Im Σ_0	T	temperature
$K^\text{I}, K^\text{II}, K^\text{III}$	off-shell contributions to Im Σ_4	T_M	Mott temperature
K_{2N}	local interaction kernel	$T_{\mu\nu}$	energy-momentum tensor
$L(\boldsymbol{x})$	Polyakov loop	τ	proper time
λ	't Hooft coupling		*or* mean free time
Λ	three-momentum cutoff	$\tau^{\mu\nu}$	dissipative tensor
m_0	current-quark mass	u^μ	four velocity
m	constituent-quark mass	v_n	flow coefficients
m_M	thermal meson mass, m_π or m_σ	v_2	elliptic flow
μ	quark chemical potential	V_a^μ	vector current of SU(N)
	or renormalization scale	ω_n	bosonic Matsubara frequencies
$n_\text{B}, n_\text{B}^\pm$	Bose distribution functions	ξ	bulk viscosity
$n_\text{F}, n_\text{F}^\pm$	Fermi distribution functions		

Bibliography

[A+06] B. Alver et al. System size and centrality dependence of charged hadron transverse momentum spectra in Au + Au and Cu + Cu collisions at $\sqrt{s_{NN}}$ = 62.4 GeV and 200 GeV. *Phys. Rev. Lett.*, 96:212301, 2006.

[A+10] K. Aamodt et al. Elliptic flow of charged particles in Pb-Pb Collisions at $\sqrt{s_{NN}}$ = 2.76 TeV. *Phys. Rev. Lett.*, 105:252302, 2010.

[Adl69] S. L. Adler. Axial-vector vertex in spinor electrodynamics. *Phys. Rev.*, 177(5):2426–2438, 1969.

[AMY00] P. B. Arnold, G. D. Moore, and L. G. Yaffe. Transport coefficients in high temperature gauge theories. 1. Leading log results. *J. High Energy Phys.*, 0011:001, 2000.

[AMY03] P. B. Arnold, G. D. Moore, and L. G. Yaffe. Transport coefficients in high temperature gauge theories. 2. Beyond leading log. *J. High Energy Phys.*, 0305:51, 2003.

[AP09] A. Armoni and A. Patella. Degeneracy between the Regge slope of mesons and baryons from supersymmetry. *J. High Energy Phys.*, 07:073, 2009.

[B+00] G. S. Bali et al. Static potentials and glueball masses from QCD simulations with Wilson sea quarks. *Phys. Rev. D*, 62:054503, 2000.

[B+06] C. A. Baker et al. An improved experimental limit on the electric dipole moment of the neutron. *Phys. Rev. Lett.*, 97:131801, 2006.

[BAA+12] N. Beisert, C. Ahn, L. F. Alday, Z. Bajnok, J. M. Drummond, et al. Review of AdS/CFT Integrability: An Overview. *Lett. Math. Phys.*, 99:3–32, 2012.

[Bal01] G. S. Bali. QCD forces and heavy quark bound states. *Phys. Rept.*, 343:1–136, 2001.

[BBS06] K. Becker, B. Becker, and J. H. Schwarz. *String Theory and M-Theory: A Modern Introduction*. Cambridge University Press, 2006.

[BDO01] N. Borghini, P. M. Dinh, and J.-Y. Ollitrault. Flow analysis from multiparticle azimuthal correlations. *Phys. Rev. C*, 64:054901, 2001.

[BHW13] N. M. Bratovic, T. Hatsuda, and W. Weise. Role of vector interaction and axial anomaly in the PNJL modeling of the QCD phase diagram. *Phys.Lett.*, B719:131–135, 2013.

[BJM87] V. Bernard, R. L. Jaffe, and U.-G. Meissner. Flavor mixing via dynamical chiral symmetry breaking. *Phys. Lett. B*, 198(1):92–98, 1987.

[BJM88] V. Bernard, R. L. Jaffe, and U.-G. Meissner. Strangeness mixing and quenching in the Nambu-Jona-Lasinio model. *Nucl. Phys. B*, 308:753–790, 1988.

Bibliography

[Bra13]   N. M. Bratovic. *Studies in a Three-Flavor PNJL Model*. Ph.D. Thesis, TUM, 2013.

[Bub05]   M. Buballa. NJL model analysis of quark matter at large density. *Phys. Rept.*, 407:205–376, 2005.

[Bur03]   E. A. Burtt. *The Metaphysical Foundations of Modern Science*. Dover Publications, 2003.

[CDVVW79] R. J. Crewther, P. Di Vecchia, G. Veneziano, and E. Witten. Chiral Estimate of the Electric Dipole Moment of the Neutron in Quantum Chromodynamics. *Phys. Lett.*, 88B:123–127, 1979.

[CEM13]   Ch. Y. Cardall, E. Endeve, and A. Mezzacappa. Conservative 3+1 general relativistic Boltzmann equation. *Phys. Rev. D*, 88:023011, 2013.

[CG73]   S. Coleman and D. J. Gross. Price of asymptotic freedom. *Phys. Rev. Lett.*, 31(13):851–854, 1973.

[CHPS14]   N. Christiansen, M. Haas, J. M. Pawlowski, and N. Strodthoff. Transport Coefficients in Yang–Mills theory and QCD. *arXiv:hep-ph/1411.7986*, 2014.

[CL06]   T.-P. Cheng and L.-F. Li. *Gauge theory of elementary particle physics*. Oxford Science Publications, 2006.

[Coh07]   T. D. Cohen. Is There a "Most Perfect Fluid" Consistent with Quantum Field Theory? *Phys. Rev. Lett.*, 99(2):021602, 2007.

[CY97]   Ch.-K. Chow and T.-M. Yan. Self-consistent $1/N_c$ expansion in the presence of electroweak interactions. *Phys. Rev. D*, 53(9):5105–5107, 1997.

[Cza05]   M. Czakon. The four-loop QCD β-function and anomalous dimensions. *Nucl. Phys. B*, 710:485–498, 2005.

[Dar00]   S. Dar. The neutron EDM in the SM: A review. *arXiv:hep-ph/0008248*, 2000.

[Def05]   H. Defu. Shear viscosity of hot QCD from transport theory and thermal field theory in real time formalism. *arXiv:hep-ph/0501284*, 2005.

[DT08]   K. Dusling and D. Teaney. Simulating elliptic flow with viscous hydrodynamics. *Phys. Rev. C*, 77:034905, 2008.

[EJ91]   S. Eidelman and F. Jegerlehner. Hadronic contributions to $(g-2)$ of the leptons and to the effective fine structure constant $\alpha(m_z^2)$. *Z. Phys. C*, 67, 1991.

[EK94]   V. L. Eletsky and I. I. Kogan. Goldberger-Treiman relation, g_A, and $g_{\pi NN}$ at $T \neq 0$. *Phys. Rev. D*, 49:R3083–R3086, 1994.

[Erd98]   R. Erdem. $1/N_c$ expansion and anomaly cancellation in the presence of electroweak interactions. *J. Phys. G*, 24:517–524, 1998.

[ESW03]   R. K. Ellis, W. J. Stirling, and B. R. Webber. *QCD and Collider Physics*. Cambridge, 2003.

[FH91]   W. Fulton and J. Harris. *Representation Theory*. Springer, 1991.

Bibliography

[FILM02] E. Ferreiro, E. Iancu, A. Leonidov, and L. D. McLerran. Nonlinear gluon evolution in the color glass condensate. 2. *Nucl. Phys. A*, 703:489–538, 2002.

[FIO] T. Fukutome, M. Iwasaki, and H. Ohnishi. Shear viscosity and spectral function of the quark matter. *arXiv:hep-ph/0606192*.

[FIO08a] T. Fukutome, M. Iwasaki, and H. Ohnishi. Effect of soft mode on shear viscosity of quark matter. *Prog. Theor. Phys.*, 119(6):991–1004, 2008.

[FIO08b] T. Fukutome, M. Iwasaki, and H. Ohnishi. Shear viscosity of the quark matter. *J. Phys. G: Nucl. Part. Phys.*, 35:035003, 2008.

[Fuj79] K. Fujikawa. Path-integral measure for gauge-invariant fermion theories. *Phys. Rev. Lett.*, 42(18):1195–1198, 1979.

[Fuk07] K. Fukushima. Sign problem in two-color two-flavor QCD with quark and isospin chemical potentials. *PoS*, LAT2007:185, 2007.

[Fuk08a] K. Fukushima. Erratum: Phase diagrams in the three-flavor Nambu-Jona-Lasinio model with the Polyakov loop. *Phys. Rev. D*, 78:039902, 2008.

[Fuk08b] K. Fukushima. Phase diagrams in the three-flavor Nambu-Jona-Lasinio model with the Polyakov loop. *Phys. Rev. D*, 77(11):114028, 2008.

[FW03] A. L. Fetter and D. J. Walecka. *Quantum Theory of Many-Particle Systems*. Dover Publications, Inc., 2003.

[GGLO12] F. G. Gardim, F. Grassi, M. Luzum, and J.-Y. Ollitrault. Mapping the hydrodynamic response to the initial geometry in heavy-ion collisions. *Phys. Rev. C*, 85:024908, 2012.

[GIJMV10] F. Gelis, E. Iancu, J. Jalilian-Marian, and R. Venugopalan. The color glass condensate. *Ann. Rev. Nucl. Part. Sci.*, 60:463–489, 2010.

[GKP96] S. S. Gubser, I. R. Klebanov, and A. W. Peet. Entropy and temperature of black 3-branes. *Phys. Rev. D*, 54:3915–3919, 1996.

[GL89] P. Gerber and H. Leutwyler. Hadrons Below the Chiral Phase Transition. *Nucl. Phys. B*, 321:387, 1989.

[GM89] C. Q. Geng and R. E. Marshak. Uniqueness of Quark and Lepton Representations in the Standard Model From the Anomalies Viewpoint. *Phys. Rev. D*, 39:693, 1989.

[GM90] C. Q. Geng and R. E. Marshak. Reply to: Comment on 'Anomaly Cancellation in the Standard Model'. *Phys. Rev. D*, 41:717–718, 1990.

[Gre05] W. Greiner. *Quantenmechanik: Symmetrien*. Verlag Harri Deutsch, 2005.

[Gro05] D. J. Gross. Nobel lecture: The discovery of asymptotic freedom and the emergence of QCD. *Rev. Mod. Phys.*, 77(3):837–849, 2005.

[Hei71] W. Heisenberg. *"Physics and Beyond: Encounters and Conversations"*. Harper & Row, 1971.

[Hel11] R. C. Helling. How I learned to stop worrying and love QFT. *arXiv:1201.2714*, 2011.

Bibliography

[HK78]   P. Hasenfratz and J. Kuti. The quark bag model. *Phys. Rept.*, 40:75–179, 1978.

[HK85]   A. Hosoya and K. Kajantie. Transport Coefficients of QCD Matter. *Nucl. Phys. B*, 250:666, 1985.

[HK87]   T. Hatsuda and T. Kunihiro. Pion and σ-meson at finite temperature. *Prog. Theor. Phys. Suppl.*, 91:284–298, 1987.

[HK94]   T. Hatsuda and T. Kunihiro. QCD phenomenology based on a chiral effective Lagrangian. *Phys. Rep.*, 247:221–367, 1994.

[HK11]   Y. Hidaka and T. Kunihiro. Computation of transport coefficients around critical point based on novel diagrammatic method. *Journal of Physics: Conference Series*, 270:012050, 2011.

[HKLW98]   M. Hess, F. Karsch, E. Laermann, and I. Wetzorke. Diquark masses from lattice QCD. *Phys. Rev. D*, 58:111502, 1998.

[HLS75]   R. Haag, J. T. Lopuszanski, and M. Sohnius. All possible generators of supersymmetries of the S-matrix. *Nucl. Phys. B*, 88:257, 1975.

[HSS12]   U. Heinz, C. Shen, and H. Song. The viscosity of quark-gluon plasma at RHIC and the LHC. *AIP Conf. Proc.*, 1441:766–770, 2012.

[HST84]   A. Hosoya, M.-a. Sakagami, and M. Takao. Nonequilibrium Thermodynamics in Field Theory: Transport Coefficients. *Annals of Physics*, 154:229–252, 1984.

[HT96]   A. Hosaka and H. Toki. Chiral bag model for the nucleon. *Phys. Rept.*, 277:65–188, 1996.

[IBY93]   N. Ishii, W. Bentz, and K. Yazaki. Faddeev approach to the nucleon in the Nambu-Jona-Lasinio (NJL) model. *Phys. Lett. B*, 301:165–169, 1993.

[ILM01]   E. Iancu, A. Leonidov, and L. D. McLerran. Nonlinear gluon evolution in the color glass condensate. 1. *Nucl. Phys. A*, 692:583–645, 2001.

[Ish98]   N. Ishii. Meson exchange contributions to the nucleon mass in the Faddeev approach to the NJL model. *Phys. Lett. B*, 431:1–7, 1998.

[Jeo93]   S. Jeon. Computing spectral densities in finite temperature field theory. *Phys. Rev. D*, 47:4586–4607, 1993.

[Jeo95]   S. Jeon. Hydrodynamic transport coefficients in relativistic scalar field theory. *Phys. Rev. D*, 52:3591–3642, Sep 1995.

[JY96]   S. Jeon and L. G. Yaffe. From quantum field theory to hydrodynamics: Transport coefficients and effective kinetic theory. *Phys. Rev. D*, 53:5799–5809, 1996.

[Kan86]   I. Kant. *Metaphysische Anfangsgründe der Naturwissenschaft*. Johann Friedrich Hartknoch, 1786.

[KG06]   J. I. Kapusta and Ch. Gale. *Finite-Temperature Field Theory*. Cambridge, 2006.

[KKM71]   M. Kobayashi, H. Kondo, and T. Maskawa. Symmetry Breaking of the Chiral $U(3) \otimes U(3)$ and the Quark Model. *Prog. Theor. Phys.*, 45(6):1955–1959, 1971.

Bibliography

[Kle92]   S. P. Klevansky. The Nambu–Jona-Lasinio model of Quantum Chromodynamics. *Rev. Mod. Phys.*, 64(3):649–708, 1992.

[KLVW90a]   S. Klimt, M. Lutz, U. Vogl, and W. Weise. Generalized SU(3) Nambu–Jona-Lasinio model (I). Mesonic modes. *Nucl. Phys. A*, 516:429–468, 1990.

[KLVW90b]   S. Klimt, M. Lutz, U. Vogl, and W. Weise. Generalized SU(3) Nambu–Jona-Lasinio model (II). From current to constituent quarks. *Nucl. Phys. A*, 516:469–495, 1990.

[Kob73]   K. Kobayashi. Eigenvalues of the Casimir operators of U(n) and SU(n) for totally symmetric and antisymmetric representations. *Prog. Theor. Phys.*, 49(1):345–347, 1973.

[Kov12]   Y. V. Kovchegov. AdS/CFT applications to relativistic heavy-ion collisions: a brief review. *Rep. Prog. Phys.*, 75:124301, 2012.

[KSS05]   P. K. Kovtun, D. T. Son, and A. O. Starinets. Viscosity in Strongly Interacting Quantum Field Theories from Black Hole Physics. *Phys. Rev. Lett.*, 94(11):111601, 2005.

[KST99]   J. B. Kogut, M. A. Stephanov, and D. Toublan. On two color QCD with baryon chemical potential. *Phys. Lett. B*, 464:183–191, 1999.

[KST$^+$00]   J. B. Kogut, M. A. Stephanov, D. Toublan, J. J. M. Verbaarschot, and A. Zhitnitsky. QCD-like theories at finite baryon density. *Nucl. Phys.*, 582:477–513, 2000.

[L$^+$07]   R. A. Lacey et al. Has the QCD critical point been signaled by observations at RHIC? *Phys. Rev. Lett.*, 98:092301, 2007.

[Lan65]   C. Lanczos. *Albert Einstein and the comsmic order*. Wiley, New York, 1965.

[Lan10]   R. Lang. *Shear Viscosity of Interacting Bose Gases*. Diploma thesis, TUM, 2010.

[LB00]   M. Le Bellac. *Thermal Field Theory*. Cambridge, 2000.

[LK96]   K. Langfeld and C. Kettner. The quark condensate in the GMOR relation. *Mod. Phys. Lett.*, 11:1331–1338, 1996.

[LKW12]   R. Lang, N. Kaiser, and W. Weise. Shear Viscosity of a Hot Pion Gas. *Eur. Phys. J. A*, 48:109, 2012.

[LP13]   K. Langfeld and J. M. Pawlowski. Two-color QCD with heavy quarks at finite densities. *Phys. Rev. D*, 88:071502, 2013.

[LR08]   M. Luzum and P. Romatschke. Conformal Relativistic Viscous Hydrodynamics: Applications to RHIC results at $\sqrt{s_{NN}} = 200$ GeV. *Phys. Rev. C*, 78:034915, 2008.

[LW13]   R. Lang and W. Weise. Shear viscosity from Kubo formalism: NJL-model study. *Eur. Phys. J. A.*, 50:63, 2013.

[Mal99]   J. Maldacena. The Large N Limit of Superconformal Field Theories and Supergravity. *Int. J. Theor. Phys.*, 38(4):1113–1133, 1999.

[Mam12]   K.-A. Mamo. Holographic RG flow of the shear viscosity to entropy density ratio in strongly coupled anisotropic plasma. *J. High Energy Physics*, 1210:070, 2012.

[Mar02]   P. Maris. Effective masses of diquarks. *Few-Body Systems*, 32:42–52, 2002.

Bibliography

[Mat76] R. D. Mattuck. *A Guide to Feynman Diagrams in the Many-Body Problem.* Dover Publications Inc., 1976.

[MBW10] D. Müller, M. Buballa, and J. Wambach. Quark propagator in the Nambu–Jona-Lasinio model in a self-consistent $1/N_c$ expansion. *Phys. Rev. D*, 81:094022, 2010.

[McE01] P. McEvoy. *Niels Bohr: Reflections on Subject and Object (Theory of Interacting Systems).* Micro Analytix, 2001.

[Mey07] H. B. Meyer. Calculation of the Shear Viscosity in SU(3) Gluodynamics. *Phys. Rev. D*, 76:101701, 2007.

[Mey08] H. B. Meyer. Calculation of the Bulk Viscosity in SU(3) Gluodynamics. *Phys. Rev. Lett.*, 100:162001, 2008.

[Moo92] W. J. Moore. *Schrödinger: Life and Thought.* Cambridge University Press, 1992.

[MRSS07] M. L. Miller, K. Reygers, S. J. Sanders, and P. Steinberg. Glauber modeling in high energy nuclear collisions. *Ann. Rev. Nucl. Part. Sci.*, 57:205–243, 2007.

[MRW90] J. A. Minahan, P. Ramond, and R. C. Warner. A Comment on Anomaly Cancellation in the Standard Model. *Phys. Rev. D*, 41:715, 1990.

[MS81] L. D. McLerran and B. Svetitsky. Quark liberation at high temperature: A Monte Carlo study of SU(2) gauge theory. *Phys. Rev. D*, 24(2):450–460, 1981.

[MS03] M. Miller and R. Snellings. Eccentricity fluctuations and its possible effect on elliptic flow measurements. *arXiv:nucl-ex/0312008*, 2003.

[Nam08] Y. Nambu. http://www.s.u-tokyo.ac.jp/en/research/alumni/nambu.html (October 1, 2014), 2008.

[NDH+11] H. Niemi, G. S. Denicol, P. Huovinen, E. Molnar, and D. H. Rischke. Influence of the shear viscosity of the quark-gluon plasma on elliptic flow in ultrarelativistic heavy-ion collisions. *Phys. Rev. Lett.*, 106:212302, 2011.

[Nes00] A. V. Nesterenko. Quark - anti-quark potential in the analytic approach to QCD. *Phys.Rev.*, D62:094028, 2000.

[NFH04] Y. Nishida, K. Fukushima, and T. Hatsuda. Thermodynamics of strong coupling two color QCD with chiral and diquark condensates. *Phys. Rept.*, 398:281–300, 2004.

[NJL61a] Y. Nambu and G. Jona-Lasinio. Dynamical model of elementary particles based on an analogy with superconductivity. I. *Phys. Rev.*, 122:345–358, 1961.

[NJL61b] Y. Nambu and G. Jona-Lasinio. Dynamical model of elementary particles based on an analogy with superconductivity. II. *Phys. Rev.*, 124:246–254, 1961.

[NS06] A. Nakamura and S. Sakai. Viscosities of hot gluon: A lattice QCD study. *Nucl. Phys. A*, 774:775–778, 2006.

[O+14] K. A. Olive et al. The Review of Particle Physics (Particle Data Group). *Chin. Phys. C*, 38:090001, 2014.

[Oll11] J.-Y. Ollitrault. Phenomenology of the little bang. *J. Phys.: Conf. Ser.*, 312:012002, 2011.

[OPV09]   J.-Y. Ollitrault, A. M. Poskanzer, and S .A. Voloshin. Effect of flow fluctuations and nonflow on elliptic flow methods. *Phys. Rev. C*, 80:014904, 2009.

[Pol73]   H. D. Politzer. Reliable Perturbative Results for Strong Interactions? *Phys. Rev. Lett.*, 30(26):1346–1349, 1973.

[PPSG12]   S. Plumari, A. Puglisi, F. Scardina, and V. Greco. Shear Viscosity of a strongly interacting system: Green-Kubo vs. Chapman-Enskog and Relaxation Time Approximation. *Phys. Rev. C*, 86:054902, 2012.

[PS95]   M. E. Peskin and D. V. Schroeder. *An Introduction to Quantum Field Theory*. Westview Press, 1995.

[PSS01]   G. Policastro, D. T. Son, and A. O. Starinets. Shear Viscosity of Strongly Coupled $N = 4$ Supersymmetric Yang-Mills Plasma. *Phys. Rev. Lett.*, 87:081601, 2001.

[QK94]   E. Quack and S. P Klevansky. Effective $1/N_c$ expansion in the NJL model. *Phys. Rev. C*, 49(6):3283–3288, 1994.

[Rö6]   S. Rößner. *Field theoretical modelling of the QCD phase diagram*. Diploma thesis, TUM, 2006.

[RS12]   A. Rebhan and D. Steineder. Violation of the Holographic Viscosity Bound in a Strongly Coupled Anisotropic Plasma. *Phys.Rev.Lett.*, 108:021601, 2012.

[RW04]   C. Ratti and W. Weise. Thermodynamics of two-colour QCD and the Nambu–Jona-Lasinio model. *Phys. Rev. D*, 70:054013, 2004.

[Sch02]   S. Scherer. Introduction to chiral perturbation theory. *Adv. Nucl. Phys.*, 27:277, 2002.

[SJG11]   B. Schenke, S. Jeon, and C. Gale. Elliptic and triangular flow in event-by-event $(3 + 1)$d viscous hydrodynamics. *Phys. Rev. Lett.*, 106:042301, 2011.

[Sne11]   R. Snellings. Elliptic Flow: A Brief Review. *New J. Phys.*, 13:055008, 2011.

[Son13]   H. Song. QGP viscosity at RHIC and the LHC - a 2012 status report. *Nucl. Phys. A*, 904-905:114c–121c, 2013.

[SR09]   C. Sasaki and K. Redlich. Bulk viscosity in quasi particle models. *Phys. Rev. C*, 79:055207, 2009.

[SR10]   C. Sasaki and K. Redlich. Transport coefficients near chiral phase transition. *Nucl. Phys. A*, 832:62–75, 2010.

[SSS01]   K. Splittorff, D. T. Son, and M. A. Stephanov. QCD-like theories at finite baryon and isospin density. *Phys. Rev. D*, 64:016003, 2001.

[Tan08]   A. Tang. Flow Results and Hints of Incomplete Thermalization. *arXiv:nucl-ex/0808.2144*, 2008.

[tH74a]   G. 't Hooft. A planar diagram theory for strong interactions. *Nucl. Phys. B*, 72:461, 1974.

[tH74b]   G. 't Hooft. A two-dimensional model for mesons. *Nucl. Phys. B*, 75:461, 1974.

Bibliography

[tH76] G. 't Hooft. Symmetry breaking through Bell-Jackiw anomalies. *Phys. Rev. Lett.*, 37(1):8–11, 1976.

[Tha06] M. Thaler. *Phases of QCD: Lattice Thermodynamics and Quasiparticle Approaches*. Ph.D. Thesis, TUM, 2006.

[Tho84] A. W. Thomas. Chiral symmetry and the bag model: A new starting point for nuclear physics. *Adv. Nucl. Phys.*, 13:1–137, 1984.

[TTM81] A. W. Thomas, S. Theberge, and G. A. Miller. The cloudy bag model of the nucleon. *Phys. Rev.*, D24:216, 1981.

[TW01] A. W. Thomas and W. Weise. *The Structure of the Nucleon*. Wiley-VCH, 2001.

[Ven10] R. Venugopalan. From many body wee parton dynamics to perfect fluid: a standard model for heavy ion collisions. *PoS*, ICHEP2010:567, 2010.

[vRVL97] T. van Ritbergen, J. A. M. Vermaseren, and S. A. Larin. The four loop beta function in quantum chromodynamics. *Phys. Lett.*, B400:379–384, 1997.

[VW91] U. Vogl and W. Weise. The Nambu and Jona-Lasinio model: Its implications for hadrons and nuclei. *Prog. Part. Nucl. Phys.*, 27:195–272, 1991.

[Wei72] S. Weinberg. *Gravitation and Cosmology: Principles and Applications of the General Theory of Relativity*. John Wiley & Sons, 1972.

[Wei99] S. Weinberg. *The Quantum Theory of Fields – Volume III*. Cambridge, 1999.

[WG73] F. Wilczek and D. J. Gross. Ultraviolet Behavior of Nonabelian Gauge Theories. *Phys. Rev. Lett.*, 30(26):1343–1346, 1973.

[Wil78] F. Wilczek. Problem of strong P and T invariance in the presence of instantons. *Phys. Rev. Lett.*, 40:279, 1978.

[Wit79] E. Witten. Baryons in the $1/N$ Expansion. *Nucl.Phys.*, B160:57, 1979.

[WLC+13] K.-l. Wang, Y.-X. Liu, L. Chang, C. D. Roberts, and S. M. Schmidt. Baryon and meson screening masses. *Phys. Rev. D*, 87(7):074038, 2013.

[WP12] A. Wiranata and M. Prakash. Shear Viscosities from the Chapman-Enskog and the Relaxation Time Approaches. *Phys. Rev. C*, 85:054908, 2012.

[YHM08] K. Yagi, T. Hatsuda, and Y. Miake. *Quark-Gluon Plasma*. Cambridge, 2008.

[ZK09] Z. Zhang and T. Kunihiro. Vector interaction, charge neutrality, and multiple chiral critical-point structures. *Phys. Rev. D*, 80:014015, 2009.

[ZSV04] L. A. P. Zayasa, J. Sonnenschein, and D. Vamanc. Regge trajectories revisited in the gauge/string correspondence. *Nucl. Phys. B*, 682(1–2), 2004.

[Zub74] D. N. Zubarev. *Nonequilibrium Statistical Thermodynamics*. Plenum NY, 1974.

Acknowledgment

"Life is wonderfully hard but you should not allow it to be hardly wonderful."

Indian saying

I am convinced that research can only succeed in an open, inspiring but also controversial environment. During my Ph.D. project I have found these conditions based on an international circle of friends and colleagues. With these final lines I try to express my gratitude to them.

In the first place, I am deeply grateful to my supervisor Prof. Dr. Wolfram Weise who gave me the opportunity to do research at his chair. Whenever I was lost in formalism he managed with his endless patience and keen insight to guide me back to the physical track. He never lost his interest in my work and helped me in countless fruitful discussions to develop the next steps. His cosmopolitan attitude allowed me to enjoy many business trips to Mainz, Washington D.C., Upton, Dresden, Wako, Kyoto, Tsukuba, Frankfurt, Darmstadt, Budapest, Heidelberg, and of course many times to Trento.

I am very thankful to my mentor Prof. Dr. Norbert Kaiser who had an open door whenever I dropped in with "unsolvable" questions. He shared his overwhelming experience and both mathematical and physical knowledge with me and supported me in so many different ways. I am grateful in particular for the chocolate supply and for providing a ride home on late Friday night when we got stuck in the office.

During my five month at RIKEN I was guided and inspired by Prof. Dr. Tetsuo Hatsuda. I have appreciated his hospitality and the opportunity to work in his group a lot. Despite his very busy schedule he was always open for physics discussions and got never tired to advice me also in touristic questions. For my perfectly organized stay in Japan I am indebted to Yoko Fujita who managed to answer an infinite number of questions. I am also thankful to Yoshimasa Hidaka for many instructive conversations and helpful advices. Also outside the office I spent many hours together with Pascal Naidon. I am grateful for this time, in particular for the invitation to his place and for the piano session in the Ohkouchi Hall only two days before my departure.

Being home in Garching it is important to find a friendly and inspiring working atmosphere. In my office this was always ensured by endless but fruitful discussions with my office mates Matthias Drews, Alexander Laschka and Thomas Hell. I am thankful to them for providing such a comfortable environment. I also thank my colleagues Stefan Petschauer, Paul Springer, and Corbinian Wellenhofer for their ability to endure my various speeches explaining physical and sometimes non-physical topics. Instructive discussions also with former members of T39 contributed to the pleasant environment and therefore to the success of this thesis. I am thankful to Michael Altenbuchinger, Nino Bratovic, Maximilian Duell, Salvatore Fiorilla, Lisheng Geng, Jeremy Holt, Kouji Kashiwa, and Betram Klein. I also thank the members and staff of the Physik Department and ECT*, in particular Susanne Tillich, Serena Avancini, and Susan Driessen.

I am obliged to my friends Domi and Steffi who suffered from many of my immature ideas and plans presented to them. It is a precious privilege to be invited to their place and to share so many hours with dancing, cooking, or going on the next quest fighting against the evil.

Most importantly I am deeply grateful to my family for their constant interest, support, and love I have experienced not only during my Ph.D. project but also during the last almost ten years since I have started my studies at TUM.

I want morebooks!

Buy your books fast and straightforward online - at one of the world's fastest growing online book stores! Environmentally sound due to Print-on-Demand technologies.

Buy your books online at
www.get-morebooks.com

Kaufen Sie Ihre Bücher schnell und unkompliziert online – auf einer der am schnellsten wachsenden Buchhandelsplattformen weltweit!
Dank Print-On-Demand umwelt- und ressourcenschonend produziert.

Bücher schneller online kaufen
www.morebooks.de

OmniScriptum Marketing DEU GmbH
Heinrich-Böcking-Str. 6-8
D - 66121 Saarbrücken
Telefax: +49 681 93 81 567-9

info@omniscriptum.com
www.omniscriptum.com

Printed by Books on Demand GmbH, Norderstedt / Germany